M. J. Schlesinger
M. G. Santoro · E. Garaci
(Eds.)

Stress Proteins

Induction and Function

With 30 Illustrations

Springer-Verlag
Berlin Heidelberg NewYork
London Paris Tokyo
Hong Kong Barcelona

Editors:

Milton J. Schlesinger
Washington University School of Medicine
Department of Molecular Microbiology
St. Louis, MO 63110, USA

M. Gabriella Santoro/Enrico Garaci
Il Universitá degli Studi di Roma
Dipartimento di Medicina Sperimentale e Scienze Biochimiche
I-00173 Roma, Italy

QP
552
.H43
S784
1990

ISBN 3-540-52776-1 Springer-Verlag Berlin Heidelberg New York
ISBN 0-387-52776-1 Springer-Verlag New York Berlin Heidelberg

Library of Congress Cataloging-in-Publication Data
Stress proteins: induction and function/Milton J. Schlesinger.
M. Gabriella Santoro, Enrico Garaci (eds.).
Includes index.
ISBN 3-540-52776-1. — ISBN 0-387-52776-1
1. Heat shock proteins. I. Schlesinger, Milton J.   II. Santoro,
M.G. (M. Gabriella)   III. Garaci, E. (Enrico)
[DNLM: 1. Heat-Shock Proteins — biosynthesis.   2. Heat-Shock
Proteins — physiology.   QU 55 S9155]
QP552.H43S784   1990
574.19'245 — dc 20

Typesetting: Best-set, Hong Kong
Offsetprinting: Mercedes Druck, Berlin; Bookbinding: Lüderitz & Bauer, Berlin
31/3020-543210 Printed on acid-free paper

Preface

The goal of much biological research in the latter half of the twentieth century has been to determine the molecular basis for a broad range of biological phenomena. Clearly, one of the most successful examples is the ability to assign the phenomena of heredity and mutation that had been studied quantitatively in whole organisms in the first half of this century to the molecular structure and replication of DNA, discovered in the 1950s. Indeed, evolutionary relationships among the diverse species in the biosphere, previously based on morphology, are now measurable in terms of the primary sequences of proteins, RNAs and DNA. There are, of course, innumerable other examples and the chapters in this volume provide one of these, namely a molecular analysis of the phenomenon initially referred to as heat shock but later found to include the more general response of cells to stress agents.

The effect of a mild heat shock on cells was first described over 25 years ago as a morphological change — a puffing — at selective sites on the polytene chromosome of *Drosophila* embryos that had been shifted from their normal temperature of 25 to 30 °C. Some 15 years later, this phenomenon could be described in terms of newly synthesized species of mRNAs and proteins. Recombinant DNA technology was in its very earliest stages of development at the time of the latter discovery, but its application to the heat shock phenomenon rapidly led not only to the cloning and sequencing of *Drosophila* heat-shock activatable genes, but to the identification of the DNA promoter sequence controlling the stress response. Studies have continued on the molecular details of heat shock gene regulation and the first chapter of this volume offers some of the most recent developments in this exciting area. Still missing from our understanding of the heat shock response is the biochemical mechanism used by a cell to "sense" the temperature shock, but we should soon have even that information as a result of in vitro studies on the interaction of the heat-shock transcription factor with DNA, as noted in data presented in Chapter 1.

One of the earliest and most profound discoveries that resulted from molecular studies of the stress response was the universality of the phenomenon. Both prokaryotes and eukaryotes respond virtually identically to a heat shock and the stress proteins themselves are strongly conserved in structure. In addition, proteins closely related to the stress proteins are found normally in cells and perform functions vital to cell growth and metabolism. Several of those functions are described in this volume. From these data we can surmise that a crucial role for several major heat-shock proteins is to protect critical cellular proteins from irreversible denaturation and to allow them to regain a native, active conformation when the stress has been removed. Another set of proteins identified as stress-induced are those normally used by all eukaryotic cells to turn over intracellular proteins via an ATP and ubiquitin-dependent proteolytic degradation pathway. The relationship between this latter system and the heat shock response is summarized in chapter 7. The important conclusion from the data and discussion of heat shock protein functions in these chapters is that protein unfolding is probably one of the most significant molecular events in the stress response. And the interplay between stress proteins and stress-induced protein unfolding could explain many facets of the phenomenon called thermo-tolerance which is described in the last chapter.

From the very first report of heat shock, it was realized that other kinds of stress could effect the temperature-induced response. The discovery of agents that can mimic a temperature shock continues and one new type, the prosta-glandins, is described here. In contrast, another chapter shows that active oxygen species, which might be expected to trigger the stress response, do not. Perhaps the most natural form of temperature stress occurs in those organisms whose life cycle involves growth under very broad temperature ranges. That the heat shock response is important to the normal life of these organisms is shown in the chapter describing two morphological forms of a fungus. The latter is but one example revealing the pervasiveness and essentiality of the stress response: truly a set of molecular events enabling organisms to survive and prosper in a biosphere ladened with environmental hazards.

Contents

Chapter 9
Mechanisms of Stress-Induced Thermo- and Chemotolerances

CHAPTER 1

Transcription of the Human HSP70 Gene: cis-Acting Elements and trans-Acting Factors Involved in Basal, Adenovirus E1A, and Stress-Induced Expression

Richard I. Morimoto, Klara Abravaya, Dick Mosser and Gregg T. Williams

Department of Biochemistry, Molecular Biology, and Cellular Biology, Northwestern University Evanston, Illinois 60208, USA

Introduction

The eukaryotic genome encodes a large multigene family of proteins that share extensive sequence identity and biochemical properties with the major heat shock-inducible protein, HSP70. In human cells there are at least five distinct members of the HSP70 gene family that are constitutively expressed or induced in response to physiological stress. The genes that encode many of these HSP70-related proteins have been cloned, and comparison of their corresponding nucleotide sequences has revealed a high degree of evolutionary conservation among members of the HSP70 family within a single species and between species (Moran et al. 1982; Lowe et al. 1983; Hunt and Morimoto 1985; Mues et al. 1986). For example, comparison of heat shock proteins between species as separated as dnaK from *E. coli* and human HSP70 reveals 50% identity at the amino acid level (Bardwell and Craig 1984). Among the human HSP70-related proteins, GRP78 and HSP70 are 76% identical while P72 and HSP70 are 85% identical (Hunt and Morimoto 1985).

The expression of HSP70 is regulated in response to diverse conditions that affect cell growth and metabolism (Fig. 1.1) ranging from entry into G1 phase of the cell cycle, during development and differentiation, and extending to acute and chronic exposure to physiological or chemically induced stress conditions (Hightower 1980; Kelley and Schlesinger 1978; Hickey and Weber 1982; Milarski and Morimoto 1986; Wu and Morimoto 1985; Banerji et al. 1986, 1987; Thomas et al. 1981; Watowich and Morimoto 1988; Shaefer et al. 1988; Mosser et al. 1988; Williams et al. 1989; Theodorakis et al. 1989). In many of these examples gene expression is rapidly induced at the transcriptional level through protein factor interactions with distinct arrays of cis-acting promoter elements located upstream of the human HSP70 gene (Hunt and Morimoto 1985; Greene et al. 1987; Wu et al. 1986a, b, 1987; Morgan et al. 1987; Mosser et al. 1988; Williams et al. 1989). Although transcriptional regulation is often the primary form of control for heat shock gene expression, regulation can also occur posttranscriptionally by the increased stability of heat shock mRNAs during heat shock (Theodorakis and Morimoto 1987) or translationally by control of both initiation and elongation of protein synthesis (DiDomenico et al. 1982; Lindquist 1980, 1981; Banerji et al. 1984, 1987; Theodorakis et al. 1988).

Stress Proteins
Schlesinger, Santoro, Garaci (Eds.)
© Springer-Verlag Berlin Heidelberg 1990

CELLULAR STRESS RESPONSE

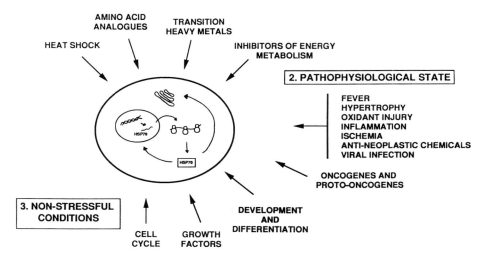

Fig. 1.1. Conditions that induce HSP70 expression

Regulation of HSP70 Expression During Cell Growth

Growth-regulated expression of the HSP70 gene was demonstrated in growth-arrested HeLa cells and a transformed human embryonic kidney cell line (293 cells) following the addition of fresh serum (Wu et al. 1985). Serum deprivation of HeLa or 293 cells resulted in reduced HSP70 transcription and mRNA levels which upon serum stimulation were induced approximately 20-fold, attaining maximal levels between 8 and 12 post-stimulation. Activation of HSP70 expression also occurs following stimulation of resting human T lymphocytes with the nonspecific mitogen phytohemagglutinin and the polypeptide growth factor interleukin-2 (IL-2) (Ferris et al. 1988). The kinetics for HSP70 expression following IL-2 stimulation of peripheral human T cells supports the growing evidence that HSP70 is a member of the growth-regulated gene family including such genes as c-myc, c-myb, and c-fos. The expression of these genes has been correlated with the transition to the growing state. In support of this, addition of the DNA synthesis inhibitor, cytosine arabinoside, during serum stimulation of HeLa cells blocks the increase in HSP70 mRNA levels. These results are consistent with the suggestion that DNA synthesis and HSP70 expression are linked (Wu and Morimoto 1985).

Changes in HSP70 mRNA levels and protein synthesis have also been observed during the cell cycle. In experiments using mitotically detached populations of synchronized HeLa cells, HSP70 mRNA levels increase at the G1/S boundary of the cell cycle in unstressed cells, and then decline rapidly during mid- to late S phase (Milarski and Morimoto 1986). The rapid increase in HSP70 mRNA levels in S-phase cells and its subsequent disappearance is due to the short half-life of HSP70 mRNA in growing cells (Theodorakis and Morimoto 1987). Coincident with the increase in message levels and HSP70 protein synthesis is the translocation of HSP70 to the nucleus of S-phase cells (Milarski and Morimoto 1986). During the G1 and G2 phase of the cell cycle, HSP70 is distributed in a diffuse pattern throughout the cytoplasm.

Transcription of the Human HSP70 Gene

Organization of the HSP70 Promoter

Comparison of the 5′-flanking sequences of HSP70 genes cloned from the *Xenopus*, mouse, rat, chicken, and human genomes have revealed a common feature of multiple cis-acting elements including one or more heat shock elements (Morgan et al. 1987; Wu B et al. 1987; Greene et al. 1987; Morimoto et al. 1986; Bienz and Pelham 1987; Williams and Morimoto 1989). Studies on human HSP70 gene expression have provided some insight into the complexity of transcriptional controls for vertebrate stress genes. Two distinct human HSP70 genes have been characterized in some detail, and each gene appears to have different transcriptional regulatory properties. The HSP70 gene isolated in Voellmy's laboratory (Voellmy et al. 1985) exhibits a less complex form of transcriptional regulation, similar to that described for many of the yeast or *Drosophila* heat-shock genes. Specifically, basal expression is low, often undetectable, and induction is in response to heat shock. A distinct HSP70 gene has been isolated and shown to be responsive to diverse cellular signals (Fig. 1.1) including serum factors, viral activation, developmental regulation, and stress induction (Greene et al. 1987; Morgan et al. 1987; Wu B et al. 1987; Williams et al. 1989). By a combination of physical mapping techniques, HSP70 genes have been mapped to chromosomes 6, 14, and 21 (Harrison et al. 1987; Sargent et al. 1989).

The promoter sequences involved in transcriptional regulation of the human HSP70 gene have been examined using transfection assays in rodent and human cell lines and by microinjection into *Xenopus* oocytes. The human HSP70 promoter is comprised of at least two functional domains (Fig. 1.2). The proximal domain which extends to −68 is required for basal expression and responds to signals that activate cell growth and transcriptional transactivation by adenovirus E1A. The distal promoter contains the heat shock element which is necessary to respond to multiple forms of stress induction. Even further upstream of −120 are additional basal elements for AP-2, CCAAT, and SP-1 that are redundant to sequences in the proximal promoter.

A

DISTAL DOMAIN   PROXIMAL DOMAIN

-120   -110   -100   -90   -80   -70   -60   -50   -40   -30   -20

GGAGGCGAAACCCCTGGAATATTCCCGACCTGGCAGCCTCATCGAGCTTCGGTGATTGGCTCAGAAGGGAAAAGGCGGGTCTCCGTGACGACTTATAAAAG(
CCTCCGCTTTGGGGACCTTATAAGGGCTGGACCGTCGGAGTAGCTCGGAGTCTTCCCTTTTCCGCCCAGAGGCACTGCTGAATATTTTCI

GAA..TTC..GAA   CCAAT   Purine   GGCGGG   TGACGTCA TATA
                         Box
HSF   CTF   Sp1   ATF   TFIID

C

		WT	Xenopus	HeLa
GCTCGGTGATTGGCTCAGAAGGGAAAAGGCGGGTCTCCGTGACGACTTATAAAAGCC		WT	1	1
GCTCGGcGAgatctcgGGGAAAAGGCGGGTCTCCGTGACGACTTATAAAAGCC		69/55	<0.1	0.1
GCTCGGTGATTGGCTCcGAgatctcgAGGCGGGTCTCCGTGACGACTTATAAAAGCC		59/49	0.3	0.3
GCTCGGTGATTGGCTCACcgaGatctcGGCGGGTCTCCGTGACGACTTATAAAAGCC G		57/47	0.7-1	0.9
GCTCGGTGATTGGCTCAGAAGGGAAAAGCCcGagaTtcgGACGACTTATAAAAGCC		44/35	0.7	0.5
GCTCGGTGATTGGCTCAGAAGGGAAAAGGCGGGTCcgaGatcgatCTcgTAAAAGCC		40/26	0.3	0.2
GCTCGGTGATTGGCTCAGAAGGGAAAAGGCGGGTCTCCGTGACcgagatctcg GCC		32/21	<0.1	0.1

B

LSN

NEO   CAT

AMP   Ori

■ Human HSP 70 Promoter Sequences
▨ Human HSP 70 Untranslated Leader Sequences
▭ Bacterial Genes
∥ Vector Sequences
□ λ Spacer DNA

Sequences Regulating Basal Expression of The HSP70 Gene

The human HSP70 promoter elements required for basal expression were identified by transfection studies in HeLa cells as well as by microinjection of promoter constructs into *Xenopus* oocytes. A collection of 5'-deletion and linker
scanner mutations were fused to a reporter gene encoding the bacterial chloramphenicol acetyl transferase (CAT). Each construct was covalently linked to a
second reporter gene encoding neomycin phosphotransferase (NEO) regulated
by the proximal domain of the HSP70 promoter, which served as a control
for transfection as well as injection efficiency (Fig. 1.2b). Analysis of 5'-deletion
mutants in *Xenopus* oocytes (K. Abravaya, unpubl. inf.) revealed that sequences
in the promoter from -74 to $+1$ were necessary and sufficient for basal expression (Fig. 1.3a). Similar results were obtained when the same 5'-deletion
mutants were assayed by transfection into HeLa cells (Williams et al. 1989).

The contribution of each cis-acting element in the basal promoter was further
examined by analysis of linker scanner mutations in two different promoter
contexts that differed according to the upstream boundary of wild-type promoter sequences. In one case HSP70 promoter sequences extended from -188
to $+1$ (LSN series), in the second case sequences extended from -100 to $+1$
(LSPN series). In either background, the wild-type promoters had the same
activity, indicating that the promoter elements between -188 and -100 do not
contribute significantly to basal activity. These results are also supported by
the results obtained with 5'-deletion mutants (Fig. 1.3a). The most drastic
effects on basal expression were observed in mutants containing substitutions for
the CCAAT and TATA elements, followed by mutations on the ATF-like and
SP1 binding sites (Fig. 1.3b, c). By comparison, alteration of the purine box
sequences did not result in a reduction of basal activity. Potential synergistic
effects or complementation of promoter sequences between -188 and -100
were noted for mutations in the CCAAT box. These results can be explained by
the presence of a distal functional CCAAT element at -160 which might be
expected to provide partial complementation. However, mutations in the ATF-
like site and the purine box were not complemented by upstream sequences.
Comparison of the relative transcriptional efficiency of the linker scanner
mutants analyzed in HeLa cells and *Xenopus* oocytes revealed that the HSP70
promoter is similarly regulated in both cell types (Fig. 1.2c).

Unlike the *Xenopus* HSP70 gene, basal expression of the human HSP70 gene
in either oocytes or HeLa cells does not depend on the heat shock element. The
Xenopus HSP70 gene is constitutively expressed in injected oocytes and heat

◁――

Fig. 1.2. Schematic of the human HSP70 promoter. A Promoter sequences from -20 to -120 are
shown. Indicated *above the sequences* are the regions corresponding to the distal and proximal domain
of the promoter. The location of transcription factor binding sites in the promoter are indicated.
B Circular map of the reporter plasmid LSN. The test promoters were introduced upstream of the
CAT gene. In opposing orientation is the NEO gene which acts as an internal standard. **C** The
nucleotide sequences of the linker scanner mutations and the phenotype of each linker scanner
mutation on basal transcription as measured by relative transcription rates in *Xenopus* oocytes and
HeLa cells

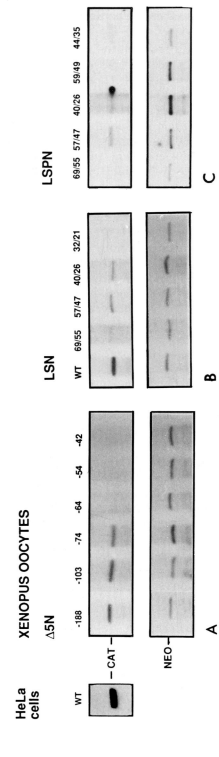

Fig. 1.3. Mutational analysis of the human HSP70 promoter function assayed in *Xenopus* oocytes. HSP70 promoter mutants fused to CAT gene as shown in Fig. 1.2 were injected into stage 6 oocytes. After 12 h incubation at 18 °C, RNA was isolated and analyzed by primer extension using 5' end labeled primers for the CAT and the NEO genes. **A** 5'deletion mutants. **B** Linker scanner mutants in the −188 to +1 promoter context: LSN series. **C** Linker scanner mutants in the −100 to +1 promoter context: LSPN series. The nomenclature is described in Fig. 1.2. Included as a standard was RNA isolated from HeLa cells transfected with the wild-type promoter fused to CAT gene

inducible in transfected somatic cells; the CCAAT box in conjunction with HSE is essential in both cases (Bienz 1984, 1986). It has been suggested that oocytes may contain high levels of activated heat shock transcription factor (HSF) that result in promoter binding together with CTF. Alternatively, an oocyte-specific form of CTF and/or HSF or a factor that stabilizes CTF-HSF interactions might exist (Bienz 1986). Our results indicate that the human HSP70 gene, when injected into *Xenopus* oocytes, does not require the HSE for basal expression, and therefore argues against the suggestion of an oocyte-specific HSF-CTF interaction. It is possible that binding sites for other transcription factors such as ATP or SP1, which are found only in the human HSP70 promoter and not in the *Xenopus* promoter, relieve the necessity for HSF in human HSP70 basal expression.

E1A Transactivation of HSP70 Gene Transcription

HSP70 transcriptional trans-activation by adenovirus E1A (Nevins 1982) has been examined by two complementary approaches (Williams et al. 1989). HeLa cells were co-transfected with a plasmid encoding E1A and HSP-CAT constructs. In addition, cell lines stably transfected with chimeric HSP70 promoters fused to the *Herpes simplex* thymidine kinase (HSP-TK) gene were infected with

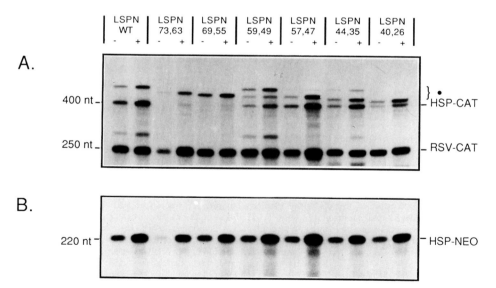

Fig. 1.4. Adenovirus E1A transactivation of the HSP70 promoter. HeLa cells were transfected with the indicated HSP-CAT constructs containing wild-type (LSPN WT) or linker scanner mutations with either the control plasmid (pAT153) or an E1A containing plasmid (pJOLC3). In addition, each set of transfections contained an E1A-noninducible negative control RSV-CAT and a positive control in the covalently linked E1A-inducible HSP-NEO construct. The size of the riboprobe-protected transcripts is indicated

adenovirus. The results of cotransfection studies using the 5'-deletion HSP-CAT mutants localized the minimum sequence required for adenovirus responsiveness to the −74 proximal domain of the HSP70 promoter. Deletion of the CCAAT box at −68 decreased both basal and E1A-inducible expression by 10- to 20-fold while mutants extending to −47 were not E1A-inducible. The use of linker-scanner mutations to examine which of the multiple elements within the HSP70 basal promoter were necessary for E1A trans-activation revealed that deletion of each element within the proximal promoter reduced the basal level of tran-scription. In each of these linker scanner mutations, the remaining promoter elements were sufficient for E1A responsiveness (Fig. 1.4). Thus E1A trans-activation was not mediated through any single promoter element but appears to activate through the entire transcription complex that assembles on sequences between −74 and the site of transcription initiation.

The role of specific HSP70 promoter elements in transcriptional regulation during adenovirus infection has also been examined using a complementary approach in which promoter elements from the HSP70 gene were inserted up-stream of the non-E1A-inducible *Herpes simplex* thymidine kinase (TK) gene. The E1A-responsiveness of each promoter construct was tested by infection of stably transfected cell lines with wild-type adenovirus. For example, fusion of a 56-nucleotide fragment, containing the HSP70 CCAAT element and flanking purine-rich sequences, to the TK gene generated a chimeric promoter (HSP-TK) which was induced during adenovirus infection with kinetics that paralleled the increase in HSP70 mRNA. Taken together, the results using transient transfec-tion and infection of stably transfected cell lines reveal that adenovirus induction of the endogenous HSP70 gene is not due to a specific chromosomal location of the HSP70 gene and that induction of the HSP70 promoter is not due to artifici-ally high levels of E1A obtained in cotransfection experiments.

cis-Acting Elements Involved in Stress-Induced Transcription

The promoter sequences necessary for heat shock and metal ion induction of the human HSP70 gene are located in the distal domain of the promoter extend-ing from −107 to −68 (Wu B et al. 1987). Within this region is a single 8/8

Fig. 1.5. Identification of the cis-acting elements required for multiple forms of stress-induced transcription. **A** S1 analysis of cytoplasmic RNA. Four plates of HeLa cells transfected with the indicated constructs were either untreated (control, *C*), incubated at 42°C for 3 h (heat shock, *HS*), incubated in the presence of 30 μM CdSO$_4$ for 7h (metal, *M*), or incubated in the presence of 5mM L-azetidine-2-carboxylic acid for 14 h (amino acid analog, *A*). S1 analysis was performed and pro-tections of HSP-CAT (*top*), HSP-NEO (*middle* band), and endogenous HSP70 (*bottom* band) transcripts are shown. **B** (*Top*) HSP70 promoter sequences from −107 to −78. Matches to the 8/8, 4/8, and 5/8 HSEs are indicated by *filled circles*, *open circles*, and *filled squares* respectively. Matches to the MRE are denoted by *open boxes*. Mismatches are indicated by an *x*. *Bottom* Schematics of mutation constructs. Each fit to the MRE is indicated above the construct. The fits to each HSE are indicated below the construct. The *thick line* denotes pBR322 sequences 2297 to 2066. **C** Sequences of mutation constructs. The relevant sequences of the promoter·mutations are shown. Mutations are depicted by *lower case letters*. The linker regions are indicated by *boldface letters*. The 8/8 HSE is *underlined*

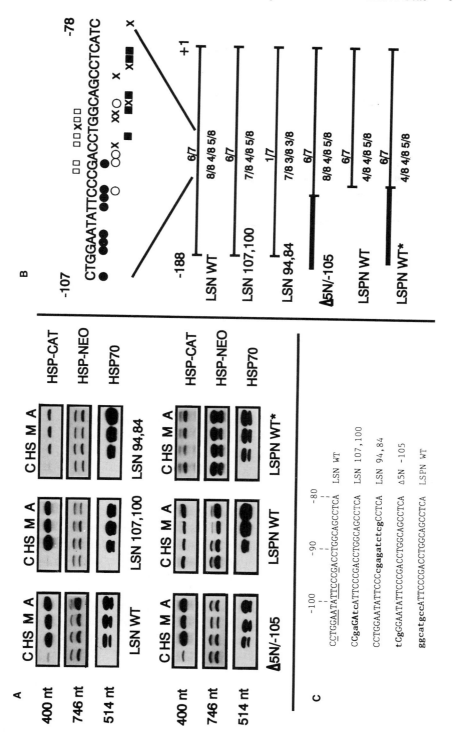

match to a 14 bp HSE consensus centered at -100 which overlaps a weak HSE (4/8 match to the consensus) centered at -90, and is adjacent to another weak HSE (5/8 match to the consensus) centered at -86 and -110 (Fig. 1.5b). Using a collection of 5'-deletion and linker-scanner mutations within these boundaries (Fig. 1.5), the minimum sequence requirement for heat shock, heavy metals, and amino acid analog activation of HSP70 transcription was shown to correspond to two inverted arrays of NGAAN centered at -100 (Williams and Morimoto 1990; Fig. 1.5c) consistent with a recent definition of the minimal functional HSE (Amin et al. 1987; Xiao and Lis 1988). Within the context of multiple cis-acting elements in the HSP70 promoter, mutations in any of the basal elements reduced both the level of basal expression and also reduced the absolute level of stress responsiveness (Williams and Morimoto 1990). Although none of the mutations in the basal promoter resulted in a complete loss of HSE-dependent stress responsiveness, the lack of a TATA element had the most severe phenotype, yet remained slightly inducible.

DNA-Protein Interactions on the Human HSP70 Promoter

The human HSP70 promoter responds to a number of cellular and environmental signals which are mediated in part through interactions between specific protein factors and a complex array of sequence elements located within 100 bp of the site of transcription initiation. As discussed, basal and E1A-inducible expression is mediated by numerous transcription factor interactions including a CCAAT element at -68, a GC element at -45, an ATF-like element at -37, and a TATA element at -28. The proteins that interact with some of these elements have been characterized by in vitro DNA binding studies and are summarized in Fig. 1.6. Many of these DNA-protein interactions can be demonstrated by performing gel mobility band shift assays using extracts from HeLa cells and a collection of double-stranded oligonucleotides that correspond to specific regions within the basal promoter. Shown in Fig. 1.6 are some of the DNA-protein complexes that are detected on low ionic strength polyacrylamide gels. The specificity of these interactions is shown by competition binding reactions in which an excess of unlabeled probe DNA has been added (lanes marked +). Further characterization including methylation interference, exonuclease III, and DNAase I protections have defined the boundaries of these DNA-protein interactions (Greene et al. 1987; Morgan et al. 1987; Wu B et al. 1987; Mosser et al. 1988).

Using the gel shift assay, a single specific interaction was detected with an oligonucleotide containing the CCAAT element and purine-rich region (Fig. 1.6, closed arrowhead). This interaction can be specifically competed by oligonucleotides containing the corresponding CCAAT elements of the HSP70 promoter, the HSV tk gene or the MSV LTR promoters. However, this interaction was not competed by a nonspecific oligonucleotide or an oligonucleotide mutated in the sites of factor interaction. From methylation interference studies the sites of interaction require contact with the two guanine residues in the CCAAT element-

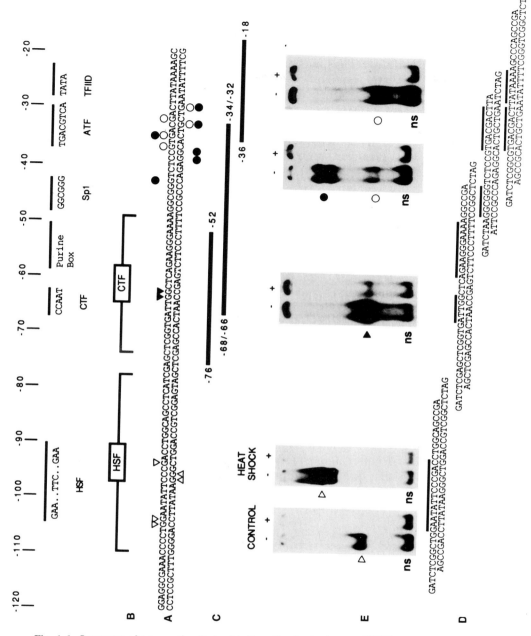

Fig. 1.6. Summary of trans-acting factor binding sites in the human HSP70 promoter. **A** The nucleotide sequences of the human HSP70 promoter and the relative position of trans-acting factor binding sites are indicated. The symbols adjacent to the sequences correspond to sites of protein contact as determined using methylation interference for HSF (*open triangles*), CTF (*closed triangles*), GC-ATF complex (*closed circles*), and ATF (*open circles*). **B** The boundaries The end-points for exonuclease III digestion are indicated. **D** Oligonucleotides used for gel mobility shift assays. Indicated as a *solid line* above each oligonucleotide sequence is the corresponding cis-acting element. **E** Gel mobility shift assays. Each set of gel shift assays corresponds to the DNA-protein complex in the absence (−) or presence (+) of self-competitor

binding site at position −68. Purified CTF will footprint the HSP70 basal promoter and can activate transcription in vitro (Morgan et al. 1987). In the −45 region lies a GC-rich region that specifically binds to affinity purified SP1. The DNAase I footprint overlaps slightly with the CTF footprint, suggesting that CTF and SP1 can bind simultaneously and perhaps interact (Morgan 1989).

Band shift assays performed with an oligonucleotide containing sequences encompassing the TATA element and the ATF-like element revealed a single specific interaction corresponding to an ATF-like factor (Fig. 1.6a, open circle). This interaction was not competed by an oligonucleotide with an altered TATA element, but was, however, competed with an oligonucleotide containing an ATF site. The sequence-specific contacts in the −35 region determined by methylation interference identified essential guanine residues within the ATF-like element that match the ATF consensus sequence in seven out of nine positions and differ from the AP1 consensus by the presence of a central cytosine residue. This suggests that the ATF-like element in the HSP70 promoter complex could interact with both ATF and AP1. Additional DNA-protein interactions were detected slightly upstream of the ATF-like element using an oligonucleotide extending from −50 to −27 of the HSP70 promoter which encompasses a GC-rich region containing the SP1 and the ATF-like element. In addition to detecting the ATF-like interactions, this oligonucleotide formed another complex of slower mobility that migrated as two closely spaced bands (Fig. 1.6, closed circle). An unusual feature of the slower-migrating DNA-protein complex is that these complexes correspond to factor interactions that are distinct from SP1 factor binding and extend over 16 nucleotides into the ATF region. Thus, it appears that the DNA-protein complex extends through the binding sites for both the SP1 and ATF sites and is distinct from both.

Given the complexity of promoters and factor-binding sites for HSP70 gene transcriptional regulation, it remains unclear which of these binding sites are essential for basal or regulated expression due to inherent limits of in vitro binding studies. A better understanding of the role of these elements will follow from studies that directly examine the binding of factors in vivo as can be obtained by genomic footprinting or DNAase I hypersensitivity experiments.

Heat Shock and Stress-Induced Transcription

Transcription Rates and HSF Levels

HeLa cells contain two heat shock element-specific DNA binding activities, one found predominantly in extracts from unstressed human cells, the other in extracts from stressed cells (Kingston et al. 1987; Sorger et al. 1987; Mosser et al. 1988). The HSE-protein complex found in extracts from unstressed cells has an electrophoretic mobility on gel shift assays distinct from that found in extracts made from stressed cells; this latter complex presumably corresponds to the heat shock transcription factor (Fig. 1.6e). Although the relationship between these two HSE-binding proteins remains unclear, both HSE-binding activities bind to DNA with nearly identical nucleotide specificities based on dimethyl sulfate interference studies. During exposure of cells to elevated temperatures, heavy metal ions, or amino acid analogs, the levels of the stress-induced activity increases. When heat-shocked cells are returned to 37 °C, the level of the stress-induced HSE-binding complex diminishes and the level of the constitutive HSE-binding activity increases. These results could be interpreted to suggest a relationship between the two HSE-binding activities; however, this remains to be rigorously demonstrated.

The HSE-protein complex formed with extracts from heat-shocked human cells has a slower electrophoretic mobility than the complex formed with extracts from unstressed cells. The stress-induced form of HSF is phosphorylated (Sorger and Pelham 1988; Larson et al. 1988). In vitro studies on human HSF (Larson et al. 1988) have led to the suggestion that phosphorylation of HSF is an essential step in transcriptional activation; however, it remains unclear whether this step is necessary for DNA binding or transcriptional activity. HSF purified from heat shock-treated human cells will function in in vitro transcription assays using extracts depleted for HSF (Goldenberg et al. 1988).

Pathways for HSF Activation

Although there is only a limited amount of information on the mechanism by which HSF activity is induced in vivo, it is likely that there are multiple pathways for HSF activation. Two inducers of the stress response, heat shock and cadmium, have different effects on HSF activation based on in vivo studies. Cells heat-shocked in the presence of cycloheximide transiently activate HSF and HSP70 gene transcription for approximately 1 h. Surprisingly, HSF is not maintained in an active state in cells continually heat-shocked in the presence of cycloheximide. By comparison, cells treated with cadmium in the presence of cycloheximide activate and maintain high levels of HSF for up to 6 h. Upon addition of cadmium to heat shock- and cycloheximide-treated cells, HSF reappears and HSP70 gene transcription is induced. These results suggest that HSF was still present in a non-DNA binding form in cycloheximide-treated heat-shocked cells. These and other results based on the use of protein synthesis inhibitors indicate that HSF is not a short-lived protein (Mosser et al. 1988; Zimarino and Wu, 1987). Additional evidence to suggest different pathways for stress-induced activation of HSF follows from the slower kinetics of HSP70 gene transcription in metal ion or amino acid analog-treated cells relative to heat-shocked cells. For example, incubation of HeLa cells with the proline analog azetidine requires 30 to 60 min to accumulate sufficient levels of de novo synthesized azetidine substituted proteins before the stress-induced HSE binding activity appears.

Characterization of the biochemical pathways by which HSF is activated in response to amino acid analog, metal ions, and heat shock will reveal firstly whether multiple pathways for stress gene activation exist and secondly how these pathways are utilized. The observation that the metal responsive element in the HSP70 promoter corresponds to the HSE has interesting implications for heavy metal transcriptional regulation. Either the HSE-binding factor is a heavy metal responsive protein distinct from metallothionein regulatory proteins that act on the human MTII gene or perhaps HSF activity is indirectly regulated by a factor that responds to intracellular flux in heavy metals. Comparison of the upstream promoter elements of the glucose responsive gene GRP78 and the metallothionein I and II genes, both of which are transcriptionally induced following incubation with cadmium, copper, and zinc (Levinson et al. 1980; Stuart et al. 1984; Karin et al. 1984; Watowich and Morimoto 1988) does not reveal a simple regulatory mechanism involving shared transcription factors. For example, the HSE is not utilized for metal-inducible transcription of either the metallothionein or GRP78 genes. Likewise, the metallothionein genes are not heat shock-responsive. Yet, because transcription of all three genes responds to the same metal ions, it is reasonable to suggest that common metal ion-sensitive intermediates in the pathway of transcription activation interact with transcription factors such as HSF.

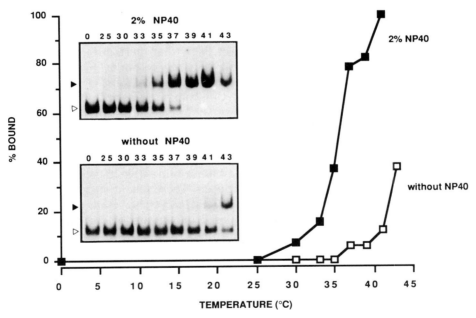

Fig. 1.7. Detergent activation of HSF reduces the temperature requirement for in vitro activation. **A** Cytoplasmic extracts were mixed with NP40 to give a final concentration of 2%, incubated at various temperatures for 60 min and analyzed by the gel mobility shift assay. Extracts heated without NP40 received an equivalent volume of water. Levels of the slower migrating complex were quantitated by scanning densitometry and are expressed as a percentage of the treatment which gave the maximum amount of binding

Biochemical Studies on HSF

Activation of HSF in vitro upon incubation of HeLa cell extracts at elevated temperatures (Larson et al. 1988) or upon treatment with low pH buffers (pH 5.8–6.4) or nonionic detergents at physiological temperatures (Mosser et al. 1990) has been obtained. In vitro studies in which extracts from unstressed cells were incubated at elevated temperatures indicate a temperature-dependent heat activation between 38 and 43 °C, with kinetics similar to that observed in vivo. HSF activity can also be activated upon treatment of S100 extracts from unstressed cells with buffers that reduce the pH, with nonionic detergents including NP40 and Triton X-100, in response to increasing concentrations of urea and in response to calcium. An important feature of all these treatments is their ability to reduce the thermal denaturation profile for HSF activation (Fig. 1.7). This allows HSF activation to occur under conditions approximating physiological temperatures instead of the extreme temperatures (greater than 43 °C) typically required to activate HSF in vitro. These studies are consistent with the suggestion that the biochemical events involved in HSF activation require altered protein conformation. Our results, however, do not distinguish between models that suggest that HSF responds to environmental signals by a direct conformational change or whether HSF responds indirectly through interactions with proteins with an altered conformation.

Acknowledgments

These studies were supported by grants from the National Institutes of General Medical Sciences, March of Dimes Foundation and an American Cancer Society Faculty Research Award (FRA313). D.M. was supported by a grant from the Medical Research Council of Canada and G.W. was supported by a NIH Training Grant in Carcinogenesis.

References

Amin J, Mestril R, Schiller P, Dreano M, Voellmy R (1987) Organization of the *Drosophila melanogaster* hsp70 heat shock regulation unit. Mol Cell Biol 7:1055–1062

Banerji SS, Theodorakis NG, Morimoto R (1984) Heat shock-induced translational control of HSP70 and globin synthesis in chicken reticulocytes. Mol Cell Biol 4:2437–2448

Banerji SS, Berg KL, Morimoto RI (1986) Transcriptional and post transcriptional regulation of avian HSP70 gene expression. J Biol Chem 261:15740–15745

Banerji SS, Laing K, Morimoto RI (1987) Erythroid lineage-specific expression and inducibility of the major heat shock protein HSP70 during avian embryogenesis. Genes and Development 1:946–953

Bardwell J, Craig E (1984) Major heat shock gene of *Drosophila* and the *Escherichia Coli* heat-inducible dnaK gene are homologous. Proc Natl Acad Sci USA 81:848–852

Bienz M (1984) *Xenopus* hsp 70 genes are constitutively expressed in injected oocytes. EMBO J 3:2477–2483

Bienz M (1986) A CCAAT box confers cell-type specific regulation on the *Xenopus* HSP70 gene in oocytes. Cell 46:1037–1042

Bienz M, Pelham H (1987) Mechanisms of heat shock gene activation in higher eukaryotes. Adv Genet 23:31–72

DiDomenico BJ, Bugaisky GE, Lindquist S (1982) The heat shock response is self regulated at both the transcriptional and posttranscriptional level. Cell 31:593–603

Ferris DK, Harel-Bellan A, Morimoto RI, Welch W, Farrar WL (1988) Mitogen and lymphokine stimulation of heat shock proteins in T lymphocytes. Proc Natl Acad Sci USA 85:3850–3854

Goldenberg G, Luo Y, Fenna M, Baler R, Weinmann R, Voellmy R, (1988) Purified human factor activates heat-shock promoter in a HeLa cell-free transcription system. J Biol Chem 263: 19734–19739

Greene, JM, Larin Z, Taylor ICA, Prentice H, Gwinn KA, Kingston RE (1987) Multiple basal elements of a human hsp70 promoter function differently in human and rodent cell lines. Mol Cell Biol 7:3646–3655

Harrison GS, Drabkin HA, Kao F-T et al. (1987) Chromosomal location of human genes encoding major heat-shock protein HSP70. Somatic Cell Mol Genet 13:119–130

Hickey ED, Weber LA (1982) Modulation of heat shock polypeptide synthesis is HeLa cells during hyperthermia and recovery. Biochemistry 21:1513–1521

Hightower LE (1980) Cultured animal cells exposed to amino acid analogs or puromycin rapidly synthesize several polypeptides. J Cell Physiol 102:407–427

Hunt C, Morimoto RI (1985) Conserved features of eukaryotic HSP70 genes revealed by comparison with the nucleotide sequence of human HSP70. Proc Natl Acad Sci USA 82:6455–6459

Karin M, Haslinger A, Holtgreve H et al. (1984) Characterization of DNA sequences through which cadmium and glucocorticoid hormones induce human metallothioniein-IIa gene. Nature (Lond) 308:513–519

Kelley PM, Schlesinger MJ (1978) The effect of amino acid analogues and heat shock on gene expression in chicken embryo fibroblasts. Cell 15:1277–1286

Kingston RE, Schuetz TJ, Larin Z (1987) Heat-inducible human factor that binds to a human hsp 70 promoter. Mol Cell Biol 7:1530–1534

Larson JS, Schuetz TJ, Kingston RE (1988) Activation in vitro of sequence specific DNA binding by a human regulatory factor. Nature (Lond) 335:372–375

Levinson W, Oppermann H, Jackson J (1980) Transition series metals and sulfhydryl reagents induce the synthesis of four proteins in eukaryotic cells. Biochem Biophys Acta 606:170–180

Lindquist S (1980) Varying patterns of protein synthesis in Drosophila during heat shock: implications for regulation. Dev Biol 77:463–479

Lindquist S (1981) Regulation of protein synthesis during heat shock. Nature (Lond) 293:311–314

Lowe DG, Fulford WD, Moran LA (1983) Mouse and Drosophila genes encoding the major heat shock protein (Hsp70) are highly conserved. Mol Cell Biol 3:1540–1543

Milarski K, Morimoto RI (1986) Expression of human HSP70 during the synthetic phase of the cell cycle. Proc Natl Acad Sci USA 83:9517–9521

Moran LA, Chawin M, Kennegy ME, Korri M, Lowe DG (1982) The major heat shock protein (hsp70) gene family: Related sequences in mouse, Drosophila, and yeast. Can J Biochem Cell Biol 51:488–499

Morgan WD (1989) Transcription factor Sp1 binds to and activates a human hsp70 promoter. Mol Cell Biol 9:4099–4104

Morgan WD, Williams GT, Morimoto RI, Greene J, Kingston RE, Tjian R (1987) Two transcriptional activators. CCAAT-box binding transcription factor and heat shock transcription factor, interact with a human hsp70 gene promoter. Mol Cell Biol 7:1129–1138

Morimoto R, Hunt C, Huang S-Y, Berg KL, Banerji SS (1986) Organization, nucleotide sequence and transcription of the chicken HSP70 gene. J Biol Chem 261:12692–12699

Mosser DD, Theodorakis NG, Morimoto RI (1988) Coordinate changes in heat shock element binding activity and HSP70 gene transcription rates in human cells. Mol Cell Biol 8:4736–4744

Mosser DD, Kotzbauer PT, Sarge K, Morimoto RI (1990) In vitro activation of heat shock transcription factor DNA/binding by calcium and biochemical conditions that affect protein conformation. Proc Natl Acad Sci USA 87:3748–3752

Mues G, Munn TZ, and Raese JD (1986) A human gene family with sequence homology to Drosophila melanogaster HSP70 heat shock genes. J Biol Chem 261:874–877

Nevins JR (1982) Induction of the synthesis of a 70 000 dalton mammalian heat shock protein by the adenovirus E1A gene product. Cell 29:913–919

Sargent CA, Dunhan I, Trowsdale J, Campbell RD, (1989) Human major histocompatibility complex contains genes for the major heat shock protein HSP70. Proc Natl Acad Sci USA 86:1968–1972

Shaefer EL, Morimoto RI, Theodorakis NG, Seidenfeld J (1988) Chemical specificity for induction of stress response genes by DNA-damaging drugs in human adenocarcinoma cells. Carcinogenesis 9:1733–1738

Sorger PK, Pelham HRB (1988) Purification and characterization of a heat shock element binding protein from yeast. EMBO J 6:3035–3041

Sorger PK, Lewis MJ, Pelham HRB (1987) Heat shock factor is regulated differently in yeast and HeLa cells. Nature (Lond) 329:81–84

Stuart GW, Searle PF, Chen HY, Brinster RL, Palmiter RD. (1984) A 12-base-pair DNA motif that is repeated several times in metallothionein gene promoters confers metal regulation to a heterologous gene. Proc Natl Acad Sci USA 87:3748–3752

Theodorakis NG, Morimoto RI (1987) Posttranscriptional regulation of hsp70 expression in human cells: Effects of heat shock, inhibition of protein synthesis, and adenovirus infection on translation and mRNA stability. Mol Cell Biol 7:4357–4368

Theodorakis NG, Banerji SS, Morimoto RI (1988) HSP70 mRNA translation in chicken reticulocytes is regulated at the level of elongation. J Biol Chem 263:14579–14585

Theodorakis NG, Zand DJ, Kotzbauer PT, Williams GT, Morimoto RI (1989) Hemin-induced transcriptional activation of the HSP70 gene during erythroid maturation in K562 cells is due to a heat shock factor-mediated stress response. Mol Cell Biol 9:3166–3173

Thomas GP, Welch WJ, Mathew MB, Feramisco JR (1981) Molecular and cellular effects of heatshock and related treatments of mammalian tissue-culture cells. Cold Spring Harbor Symp, Vol XLVI Cold Spring Harbor, NY, pp 985–996

Voellmy R, Ahmed A, Schiller P, Bromley P, Rungger D (1985) Isolation and functional analysis of a human 70 000 dalton heat shock protein gene segment. Proc Natl Acad Sci USA 82:4949–4953

Watowich SS, Morimoto RI (1988) Complex regulation of heat shock and glucose responsive genes in human cells. Mol Cell Biol 8:393–405

Williams GT, Morimoto RI (1990) Maximal stress-induced transcription requires interaction with basal promoter element upstream of the human HSP70 gene independent of rotational alignment. Mol Cell Biol 10:3125–3136

Williams GT, McClanahan TK, Morimoto RI (1989) E1A-Transactivation of the human HSP70 promoter is mediated through the basal transcription complex. Mol Cell Biol 9:2574–2587

Wu BJ, Morimoto RI (1985) Transcription of the human hsp70 gene is induced by serum stimulation. Proc Natl Acad Sci USA 82:6070–6074

Wu BJ, Hunt C, Morimoto RI (1985) Structure and expression of the human gene encoding major heat shock protein HSP70. Mol Cell Biol 5:330–341

Wu BJ, Kingston RE, RI Morimoto (1986a) Human HSP70 promoter contains at least two distinct regulatory domains. Proc Natl Acad Sci USA 83:629–633

Wu BJ, Hurst HC, Jones NC, Morimoto RI (1986b) The E1A 13S product of adenovirus 5 activates transcription of the cellular human HSP70 gene. Mol Cell Biol 6:2994–2999

Wu BJ, Williams GT, RI Morimoto (1987) Detection of three protein binding sites in the serumregulated promoter of the human gene encoding the 70-kDa heat shock protein. Proc Natl Acad Sci USA 84:2203–2207

Wu C, (1984) Activating protein factor binds in vitro to upstream control sequences in heat shock gene chromatin. Nature (Lond) 311:81–84

Wu C, Wilson S, Walker B, Dawid I, Paisley T, Zimarino V, Ueda H, (1987) Purification and properties of *Drosophila* heat shock activator protein. Science 238:1247–1253

Xiao H, Lis JT (1988) Germline transformation used to define key features of heat-shock response element. Science 239:1139–1142

Zimarino V, Wu C (1987) Induction of sequence-specific binding of *Drosophila* heat shock activator protein without protein synthesis. Nature (Lond) 327:727–730

CHAPTER 2

Active Oxygen Species and Heat Shock Protein Induction

Roy H. Burdon[1], Vera Gill[1] and Catherine Rice Evans[2]

[1] Department of Bioscience and Biotechnology,
   Todd Centre, University of Strathclyde, Glasgow G4 ONR, Scotland
[2] Department of Biochemistry, Royal Free Hospital School of Medicine, London NW3, 2PF,
   England

Introduction

As part of the cellular response to sublethal heat stress there is a rapid and co-ordinated increase in the expression of a group of proteins, the heat shock proteins (HSPs), or stress proteins (see Burdon 1986, 1988; Lindquist 1986). It is now clear that at least some of these stress proteins are essential for the survival of cells confronted with temperature and other stresses. In addition to elevated temperatures, a variety of other treatments will elicit the increased production of the HSPs. These include certain heavy metals, high levels of ethanol and amino acid analogues. Because of the possibility that these alternative inducers, like heat, could create abnormal protein structures within the cell, it is believed that the intracellular accumulation of abnormal proteins may be an important element in mechanisms leading to HSP induction (Munro and Pelham 1985; Anathan et al. 1986).

HSP induction has also been observed in insect cells (*Drosophila*) and in mammalian cells (CHO) reoxygenated following a period of anoxia (Ropp et al. 1983; Li and Shrieve 1982). This led to speculation that superoxide radicals might also be HSP inducers (Ropp et al. 1983). Many kinds of tissue and cellular stresses are known to result in an increased production of oxygen-derived free radicals and other active oxygen species. Besides ischaemia or anoxia, these include inflammatory responses, drug toxicity and X-irradiation (see Halliwell and Gutteridge 1985). Together, these observations raise the question of active oxygen production during thermal injury to cells and whether this has significance for HSP induction.

The stressful effects of oxygen towards normal cells are not due to its own reactivity, which is rather feeble, but to its ability to undergo a series of one-electron reduction processes. Uni-electron reduction of oxygen generates superoxide radicals, $O_2^{\cdot-}$. Normal cells, however, are equipped with the enzymes Cu/Zn-superoxide dismutase (Cu/Zn-SOD) and Mn-superoxide dismutase (Mn-SOD) which will catalyse the conversion of superoxide radicals to hydrogen peroxide. This hydrogen peroxide is normally metabolised intracellularly through two enzyme systems. These are catalase and the selenium-containing glutathione-dependent enzyme glutathione peroxidase (GSH-Px).

Stress Proteins
Schlesinger, Santoro, Garaci (Eds.)
© Springer-Verlag Berlin Heidelberg 1990

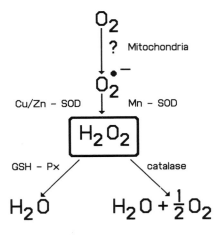

Fig. 2.1 Interrelationships between molecular oxygen and certain active oxygen species

Superoxide Radicals and HSP Induction

In preliminary studies it was possible to detect a low but steady production of superoxide radicals within cultured HeLa cells even at normal growth temperatures (Burdon et al. 1990). MTT, a tetrazolium-based compound, permeates cells and is reduced by superoxide radicals to give a blue formazan which is extractable with dimethyl sulphoxide, (DMSO) (see Carmichael et al. 1987 and Table 2.1). Although a possible source of these superoxide radicals is the enzyme xanthine oxidase (Eddy et al. 1987), there is no change in rate of production if the xanthine oxidase inhibitor allopurinol is included in the culture medium, at 50 μM. A more likely source is mitochondria (Boveris 1977; Nohl et al. 1978). In normal cells, whilst almost all molecular oxygen is fully reduced by cytochrome oxidase, it is recognised that electrons can "leak" inappropriately from the respiratory chain, with the resultant production of reduced oxygen species such as superoxide rather than water.

From the results in Table 2.1 it can be seen that the application of thermal stress at 42° or 45°C can cause increased superoxide production in HeLa cells. Moreover, there is also an apparent leakage of superoxide from these cells into the medium at the elevated temperatures which is not detectable at 37°C. This leakage could be an outcome of structural changes in cell membranes which are known to follow hyperthermic exposure (Lee and Chapman 1987). Table 2.1 also shows that the addition of diethyldithiocarbamate (DDC), an inhibitor of Cu/Zn-SOD, along with MTT leads to an even greater production of blue formazan at both normal and elevated temperatures. This ability of DDC to increase the level of superoxide reacting with MTT raises the question whether addition of DDC, by allowing possible intracellular superoxide accumulation, would also serve to induce HSP synthesis. However, Fig. 2.2 shows that although DDC addition did not interfere with heat induction of HSP70 group proteins, it did not serve as an inducer on its own.

Table 2.1. MTT-formazan production in cultures of HeLa cells subject to hyperthermia and the effect of diethyldithiocarbamate (DDC).

Expt	Conditions	Blue formazan accumulation (A_{570}) Intracellular	Extracellular
1	37 °C, 30 min	0.011 ± 0.006	None detected
	42 °C, 30 min	0.064 ± 0.006	0.218 ± 0.042
	45 °C, 30 min	0.084 ± 0.018	0.281 ± 0.071
2	37 °C, 120 min	0.210 ± 0.015	Not determined
	37 °C, 120 min + 1 mM DDC	0.465 ± 0.050	"
	45 °C, 20 min	0.255 ± 0.060	"
	45 °C, 20 min + 1 mM DDC	0.875 ± 0.025	"

Triplicate cultures of HeLa cells (0.5×10^6 cells per 3.5-cm Petri dish) were established as monolayers in 2 ml Glasgow modification of Eagle's minimal essential medium supplemented with 10% calf serum by growth at 37 °C for 24 h. The medium was then replaced, 0.25 mg MTT [3-(4, 5-dimethylthiazol-2-yl)-2,5-diphenyltetrazolium bromide] added (zero time) and incubation continued at 37, 42, or 45 °C.
At the indicated times the medium was removed and the absorbance of the blue formazan in the medium determined at 570 nm (extracellular). The remaining cell monolayers were then extracted with 2 ml DMSO and the absorbance of the intracellular blue formazan determined at 570 nm. Results represent the means of formazan absorbance from triplicate cultures $\pm$ S.D.

How Important is Hydrogen Peroxide?

Whilst the above experiments do not rule out a role for superoxide radicals in HSP induction, the possibility that other active oxygen species may be important was examined. Superoxide radicals, of course, can either dismutate spontaneously or in a reaction catalysed by SOD to yield hydrogen peroxide (McCord and Fridovich, 1988). Importantly, addition of exogenous hydrogen peroxide to cultured hamster fibroblasts (Spitz et al. 1987) will bring about elevated HSP production. In HeLa cells we find that relatively high concentrations of hydrogen peroxide of between 0.1 and 0.5 mM were required (Burdon et al.

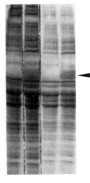

Fig. 2.2 Fluorogram of an SDS/polyacrylamide gel of HeLa cell proteins following treatment of HeLa cells with DDC in the presence or absence of hyperthermia. HeLa cells were exposed to ^{35}S-methionine (Burdon et al. 1982) following various experimental procedures. The labelled proteins were isolated and analysed by electrophoresis as described in Burdon et al. (1982). Track *1*, untreated cells; track *2*, cells treated for 10 min at 45 °C and then returned to 37 °C for 2 h; track 3, cells treated for 2 hr at 37 °C with 1 mM DDC; track *4*, cells treated with 1 mM DDC for 2 h at 37 °C, then held at 45 °C for 10 min followed by return to 37 °C for 2 h. The arrowhead indicates the electrophoretic migration of HSPs of the 70 kDa group

1 2 3 4

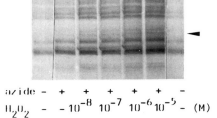

aʑide – +   +   +   +   +   –

H_2O_2 –   – 10^{-8} 10^{-7} 10^{-6} 10^{-5} –   (M)

Fig. 2.3 Fluorogram of an SDS/polyacrylamide gel of HeLa cell proteins following treatment of HeLa cells with H_2O_2 in the presence of sodium azide. Analytical procedures were as described in Fig. 2.2. Sodium azide was added to cell culture medium where indicated at 5 mM and the concentrations of H_2O_2 used are detailed below each track. The *arrowhead* indicates the electrophoretic migration of HSPs of the 70 kDa group

1990). On the other hand, if sodium azide (inhibitor of catalase) is added along with hydrogen peroxide, then concentrations of exogenous hydrogen peroxide as low as 0.001 mM were effective (Fig. 2.3).

Whilst hydrogen peroxide appears to have HSP-inducing capability, it is also able to stimulate overall cellular protein synthesis. For example, low concentrations have a stimulatory effect in vitro when added to cell-free protein synthesis systems from HeLa cells (Burdon et al. 1990). Moreover, when hydrogen peroxide was added to cultures of HeLa cells in the presence of azide there was also a notable stimulatory effect on cellular protein synthesis (Burdon et al. 1990). This effect is evident from Fig. 2.3, where there is an overall increase in the extent of labelling of cell proteins of HeLa cells when both hydrogen peroxide and azide are added to the culture medium.

Although exogenous hydrogen peroxide can induce HSPs in cultured mammalian cells in the presence or absence of the catalase inhibitor sodium azide, a question is whether hydrogen peroxide derived intracellularly from superoxide radicals can contribute to HSP induction. To approach this problem, cultures of HeLa cells were exposed to MTT during both the period of exposure to elevated temperature and the recovery period at 37°C. The MTT would be expected to consume superoxide radicals produced during the period of heat shock and so lower the potential for intracellular hydrogen peroxide production. From Fig. 2.4 it appears that such treatment reduces the heat induction of at

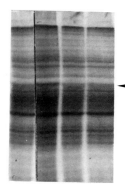

1   2   3   4

Fig. 2.4 Fluorogram of HeLa cell proteins following heat shock and recovery in the presence of MTT. Analytical procedures were as described in Fig. 2.2. Track *1* untreated cells; track *2* cell heated at 45 °C for 10 min, then returned to 37 °C for 2 h; track *3* cells treated with MTT (0.25 mg/plate) for 2 h; track *4* cells treated as for track 3 but then heated at 45 °C for 10 min and allowed to recover at 37 °C for 2 h without removal of MTT. The *arrowhead* indicates the electrophoretic migration of HSPs of the 70 kDa group

least the HSP70 group of proteins. Thus the intracellular generation of H_2O_2 from superoxide during heat shock may be a significant contributory factor in HSP induction following hyperthermia.

Hydroxyl Radicals and HSP Induction

It has often been suggested that hydrogen peroxide may have a role in cell killing through the formation of hydroxyl radicals (see Halliwell and Gutteridge 1985). Experiments were carried out in which HeLa cells were *both* heat shocked (45°C, 10 min) *and* allowed to recover (2 h, 37 °C) in the continuous presence of a variety of hydroxyl radical scavengers. The presence of 5 mM mannitol, 0.1% (v/v) dimethyl sulphoxide, 5 mM sodium formate, or 5 mM thiourea, however, had little significant effect in depressing HSP induction by heat, suggesting that hydroxyl radicals may not be significantly contributory to HSP induction mechanisms.

Intracellular Superoxide, Hydrogen Peroxide and the Abnormal Protein Hypothesis of HSP Induction

Although the experiments reported here suggest that hydrogen peroxide derived from superoxide radicals may be the most critical active oxygen species in relation to HSP induction, a question relates to its significance in light of hypotheses that abnormal proteins are important components of the HSP induction mechanisms (Anathan et al. 1986). Recent support for the abnormal protein hypothesis arises out of experiments of Edington et al. (1989), who demonstrated that HSP induction can be suppressed if the cells are exposed to heat in the presence of deuterium oxide or glycerol. Both of these agents are known to have a stabilising effect on cellular proteins under the conditions of heat stress employed (Lin et al. 1984).

From Fig. 2.5 it will be seen that when HeLa cells are subject to elevated temperature stress, but in the presence of added glycerol, the expected elevated level of superoxide production is much reduced. In turn this would limit hydrogen peroxide formation. Thus although heat-induced destabilisation of proteins may be important *per se* in HSP triggering mechanisms, such destabilisation (reduced in the presence of glycerol) may also affect mitochondrial integrity (Bowler 1987). This might result in increased electron leakage from the electron-transport chain, which might cause an ephemeral elevation of intracellular hydrogen peroxide levels and HSP induction. On the other hand, it must be conceded that hydrogen peroxide itself could induce abnormal protein structures within cells due to specific oxidation of protein sulphydryl groups. An alternative is that although superoxide radicals might give rise to short-term elevations in hydrogen peroxide levels through heat-induced destabilisation of

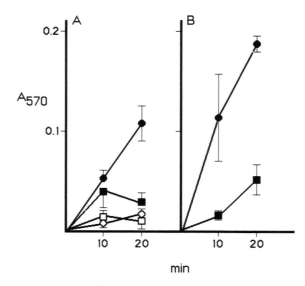

Fig. 2.5 Effect of glycerol on MTT-formazan production by HeLa cells. The experimental procedure described in Table 1 was adopted and the accumulated MTT-formazan was determined (**A**) intracellularly and (**B**) extracellularly at the times indicated. Where specified, glycerol was added at 20% (*v/v*). Results represent the means of absorbance due to formazan at 570 nm from triplicate cell cultures ± S.D. Cells at 37 °C (0); cells at 37 °C plus gylcerol (□); cells at 45 °C (●), cells at 45 °C plus glycerol (■)

mitochondrial systems, the hydrogen peroxide may itself directly activate the heat shock transcription factors (Pelham 1985) necessary for the transcription of HSP genes.

Summary

Whilst an increase in the intracellular production of superoxide radicals is detectable within cultured HeLa cells following heat shocks, studies with diethydithiocarbamate, inhibitor of Cu/Zn superoxide dismutase, did not indicate a direct role for superoxide radicals in heat shock protein induction. Likewise, experiments carried out with hydroxyl radical scavengers did not suggest a function for hydroxyl radicals in the induction mechanism. Rather it appears that hydrogen peroxide, which can be derived intracellularly from superoxide radicals, may be the critical active oxygen species in relation to heat shock protein induction.

References

Anathan J, Goldberg AC, Voellmy R (1986) Abnormal proteins serve as eukaryotic stress signals and trigger activation of heat shock genes. Science 232:522–524

Boveris A (1977) Mitochondrial production of superoxide radical and hydrogen peroxide. Adv Exp Med Biol 78:67–75

Bowler K (1987) Cellular heat injury: are membranes involved? In: Bowler K, Fuller BJ (eds) Temperature and animal cells. Symp Soc Exp Biol 41:157–185, The Company of Biologists Ltd, Cambridge

Burdon RH (1986) Heat shock and the heat shock proteins. Biochem J 240:313–324

Burdon RH (1988) The heat shock proteins. Endeavour 12:133–138

Burdon RH, Slater A, McMahon M, Cato ACB (1982) Hyperthermia and the heat shock proteins of HeLa cells. Br J Cancer 75:953–963

Burdon RH, Gill V, Rice-Evans C (1990) Radicals and stress protein induction In: Rice-Evans C et al. (eds) Stress Proteins in inflammation. Richelieu Press, Lond (in press)

Carmichael J, De Graff WG, Gazdar AF, Minna JD, Mitchell JB (1987) Evaluation of a tetrazolium-based semiautomated colorimetric assay: assessment of chemosensitivity testing. Cancer Res 47:936–942

Eddy LJ, Stewart JR, Jones HP, Engerson TD, McCord JM, Downey JM (1987) Free radical producing enzyme, xanthine oxidase is undetectable in human hearts. Ann J Physiol 253 H-709

Edington BV, Whelan SA, Hightower LE (1989) Inhibition of heat shock stress protein induction by deuterium oxide and glycerol: additional support for the abnormal protein hypothesis of induction. J Cell Physiol 139:219–228

Halliwell B, Gutteridge JMC (1985) Free radicals in biology and medicine. Clarendon Press, Oxford

Lee DC, Chapman D (1987) The effects of temperature on biological membranes and their models. In: Bowler K, Fuller BJ (eds) Temperature and animal cells. Symp Soc Exp Biol 41:35–52. The Company of Biologists Ltd, Cambridge

Li GC, Shrieve DC (1982) Thermal tolerance and specific protein synthesis in Chinese hamster fibroblasts exposed to prolonged hypoxia. Exp Cell Res 142:464–468

Lin P-S, Hefter K, Ho K-C (1984) Modification of membrane function, protein synthesis and heat killing effect in cultured Chinese hamster cells by glycerol, and D_2O. Cancer Res 44:5776–5784

Lindquist S (1986) The heat shock response. Ann Rev Biochem 55:1151–1191

McCord JM, Fridovich I (1988) Superoxide dismutase: the first twenty years, 1968–1988. Free Rad Biol Med 5:363–369

Munro S, Pelham H (1985) What turns on heat shock genes? Nature (Lond) 137:477–478

Nohl H, Breuninger V, Henger D (1978) Influence of mitochondrial radical formation on energy linked respiration. Eur J Biochem 90:385–393

Pelham HRB (1985) Activation of heat shock genes in eukaryotes. Trends Genet 1:31–35

Ropp M, Courgeon AM, Calvayrae R, Best-Belpomme (1983) The possible role of the superoxide ion in the induction of heat shock and specific proteins in aerobic *Drosophila* cells after a period of anaerobiosis. Can J Biochem Cell Biol 61:456–461

Spitz DR, Dewey WC, Li GC (1987) Hydrogen peroxide or heat shock induces resistance to hydrogen peroxide in Chinese hamster fibroblasts. J Cell Physiol 131:364–373

CHAPTER 3

Induction of HSP70 by Prostaglandins

M. Gabriella Santoro[1,2], Enrico Garaci[1] and Carla Amici[1]

[1] Department of Experimental Medicine, II University of Rome, Tor Vergata,
[2] Institute of Experimental Medicine, National Council of Research, Rome, Italy

Introduction

Prostaglandins (PGs) are a class of naturally occurring cyclic 20-carbon fatty acids, synthesized by most types of eukaryotic cells from polyunsaturated fatty acid precursors. Arachidonic acid (eicosatetraenoic acid) is the major source of PGs in mammalian tissue, together with eicosatrienoic and eicosapentaenoic acids from the phospholipid pool of the cell membrane, from which they are released in response to external stimuli through the action of phospholipase A and a phosphatidylinositol-specific phospholipase C (reviewed by Samuelsson 1982). PGs are designated by groups, depending on the structure of the cyclopentane ring, and are characterized as mono-, bi- and tri-unsaturated compounds, depending on the number of C-C double bonds in the aliphatic side chains.

Since their discovery, they have been shown to function as microenvironmental hormones and intracellular signal mediators and to participate in the regulation of a large variety of physiological and pathological processes, including cell proliferation and differentiation (Garaci et al. 1987), the immune response (Ninnemann 1988), inflammation (Vane 1987), cytoprotection (Robert 1981; Ruwart et al. 1981), and the pathology of fever (Dinarello and Wolff 1982).

Starting from the first observations that prostaglandin E_1 could inhibit the growth of L-929 mouse fibroblast cells (Johnson and Pastan 1971) and that prostaglandins of the E type suppress the proliferation of mouse B16 melanoma cells in vitro as well as the growth of B16 melanoma tumors in vivo (Santoro et al. 1976), an increasing amount of literature has described the antiproliferative activity of several prostaglandins in a large number of experimental models in vitro and in vivo (Bregman et al. 1986; Shahabi et al. 1987; Garaci et al. 1987; D'Onofrio et al. 1990a; Marini et al. 1990). However, the mechanism by which selected PGs can control cell proliferation is still mainly unknown.

For several years we have studied the role of prostaglandins in the proliferation and differentiation of tumor cells. One of our objectives was to identify at the molecular level some of the potential sites of action of prostaglandins, as well as the metabolic events associated with the antiproliferative activity of PGs. Because of the lack of data in this area and the complexity of

Stress Proteins
Schlesinger, Santoro, Garaci (Eds.)
© Springer-Verlag Berlin Heidelberg 1990

eukaryotic cell metabolism, we started by analyzing simpler, well-defined models, like viruses, to study the molecular events that follow prostaglandin administration. This approach not only gave us useful information on PG-mediated alterations of macromolecular metabolism, but led us to the interesting discovery that prostaglandins of the A and J type are potent inhibitors of virus replication (Santoro et al. 1980b). One of the first observations that we obtained studying the effect of PG treatment on protein synthesis of an African green monkey kidney (AGMK) cell line infected with Sendai virus was the accumulation of a cellular protein of 74 kDa mol. wt., that we called p74, after PGA_1 treatment (Santoro et al. 1982b). The interest in the identification of this protein increased when, a few years later, we found that the synthesis of a polypeptide of the same molecular weight was induced by PGA_1 in a different cell line, the human K562 erythroleukemia (Santoro et al. 1986). On the basis of the molecular weight and other biochemical characteristics, we hypothesized that p74 could be a stress protein (Santoro 1987). This protein has, in fact, been recently identified as a heat shock protein (HSP), related to the major HSP70 group (Santoro et al. 1989a); for the purpose of this report we will refer to it as p74.

In this chapter we summarize the results of our recent work on the antiproliferative and the antiviral action of prostaglandins, which suggest that induction of heat shock proteins could be associated with both these activities of PGs.

Induction of HSP70 During the Prostaglandin-Induced Block of Cell Proliferation

Prostaglandins of the A-, D-, and E-series have been shown to inhibit cell proliferation and/or promote cell differentiation in many tumor models (Garaci et al. 1987; Fukushima et al. 1989) among which are several murine and human leukemic cell lines, which include the Friend erythroleukemia (Santoro et al. 1980a; Marini et al. 1990), WEHI-3B-D⁻ myelomonocytic leukemia (Moore 1982), MI mouse myeloid leukemia (Honma et al. 1980), L-1210 leukemia (Narumiya and Fukushima 1985), HL-60 human promyelocitic leukemia (Breitman 1987) and U-937 human lymphoma (Olsson et al. 1982).

Types A and J prostaglandins, which are characterized by the presence of an α, β-unsaturated carbonyl group in the cyclopentane ring (cyclopentenone PGs), and are produced by enzymatic dehydration of PGEs and PGDs, respectively (Narumiya and Fukushima 1985; Ohno et al. 1986), are the most active in controlling cell proliferation and have been shown to arrest cells in the G_1 phase of the cell cycle (Hughes-Fulford 1987; Bhuyan et al. 1986; Ohno et al. 1988).

Cyclopentenone prostaglandins have been shown to be actively and selectively transported into cells by a carrier on the cell membrane. They are then transferred to the nuclei where they bind to nuclear proteins (Narumiya and

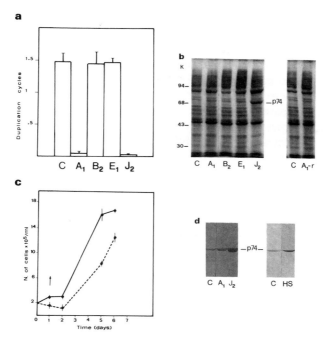

Fig 3.1. Effect of different PGs on cell proliferation and HSP70 synthesis in K562 cells. K562 cells, plated at the density of 2×10^5 cells/ml in RPMI 1640 medium, were treated with 4 µg/ml of PGA$_1$ (*A$_1$*), PGB$_2$ (*B$_2$*), PGE$_1$ (*E$_1$*), PGJ$_2$ (*J$_2$*) or ethanol diluent (*C*) at the time of plating (for culture conditions see Santoro et al. 1986). Cell numbers were determined 48 h after plating (**a**). SDS-PAGE analysis of ^{35}S-methionine-labeled (10 µCi/10^6 cells for 24 h) polypeptides from control (*C*) and PG-treated K562 cells (4 µg/ml for 24 h) shows the synthesis of a 74 kDa protein, p74, in cells treated with PGA$_1$ and PGJ$_2$ (**b**). **c** K562 cells treated with PGA$_1$ for 24 h were washed three times and replated with fresh medium devoid of PGA$_1$ (*arrow*). After 4 days, PGA$_1$-pretreated cells (*A$_1$-r*) were labeled with ^{35}S-methionine (10 µCi/10^6 cells for 24 h) and samples were analyzed by SDS-PAGE. At this time p74 was not detected (**b**). **d** Identification of p74 as a HSP70 by immunoblot analysis. Equal amounts of proteins from control (37°C) (lane *C*) or heat-shocked (45°C for 10 min) (lane *HS*) cells, and from K562 cells treated with PGA$_1$ (*A$_1$*), PGJ$_2$ (*J$_2$*), or ethanol diluent (*C*) were separated by SDS-PAGE. Immunoblot analysis, using a monoclonal anti 72/73 kDa HSP antibody from HeLa cells, revealed a 74 kDa protein whose synthesis was increased severalfold in heat-shocked (HS)-, PGA$_1$ (A$_1$)-, and PGJ$_2$ (J$_2$)-treated cells as compared to control K562 cells (*C*). (For technical details see Santoro et al. 1989a)

Fukushima 1986; Narumiya et al. 1986) and act on a specific phase of the cell cycle, at the G$_1$/S interface (Hughes-Fulford 1987; Ohno et al. 1988). Cellular uptake was found to be closely correlated to growth inhibition (Narumiya et al. 1987). In contrast with many other aspects of prostaglandins research, the antiproliferative activity of cyclopentenone PGs appears to be independent of changes in cAMP intracellular concentration (Hughes-Fulford et al. 1985).

The K562 cell line, established by Lozzio and Lozzio (1975) from a patient with chronic myelogenous leukemia, was shown to be extremely sensitive to the antiproliferative action of cyclopentenone PGs (Fig. 3.1a). When K562 cells were treated with identical concentrations of prostaglandins of the A, B, E, F, and J type, only PGA$_1$ and PGJ$_2$ totally suppressed cell proliferation without affecting cell viability (Santoro et al. 1986, 1989a). PGA_1 antiproliferative activity was reversible depending on the length of treatment, since removal of PGA_1 after 96 h resulted in a full recovery of growth potential, while, if a second 96-h treatment was started, K562 cell proliferation appeared to be definitively impaired, and it was not restored upon removal of PGA_1; moreover inhibition was not due directly to suppression of DNA synthesis, which was not altered for at least 24 h after treatment and was accompanied by a slight (10 to 30%) inhibition of protein synthesis and glycosylation (Santoro et al. 1986).

In order to investigate whether inhibition of cell replication was associated with alteration of protein metabolism, K562 cells were treated with PGA_1, PGB_2, PGE_1, PGJ_2 (9-deoxy-Δ^9-PGD_2) or ethanol diluent and the SDS-PAGE profiles of ^{35}S-methionine-labeled proteins were analyzed 24 h after treatment (Fig. 3.1a, b). A 74 kDa protein (p74) accumulated in PGA_1-treated and was even more pronounced in PGJ_2-treated cells, whose proliferation was almost completely suppressed for up to 4 days after plating (Santoro et al. 1989a). PGs that did not inhibit cell proliferation, PGB_2 and PGE_1 (as well as PGE_2 and $PGF_{1\alpha}$, data not shown), did not produce any significant change in protein metabolism at the dose used and did not induce p74 synthesis (Fig. 3.1a, b). Moreover, if K562 cells were released from the PGA_1-induced proliferation block, p74 was no longer produced when cells had regained their full growth potential (Fig. 3.1b, c).

The PG-induced p74 is not glycosylated and has a pI value of ~5.4. Its synthesis starts 1–2 h after PG treatment and is inhibited by actinomycin D treatment (unpubl. results). On the basis of its molecular mass and other biochemical characteristics similar to the major human 72/74 kDa HSP (Burdon 1982), we hypothesized that p74 could be a stress protein, synthesized by the cell in response to a stress situation (Santoro 1987).

p74 has now been identified as a heat shock protein related to the HSP70 group (Santoro et al. 1989a). Immunoblot analysis of cell extracts from control, PGA_1-treated, PGJ_2-treated and heat-shocked K562 cells using monoclonal antibodies against the human 72/73 kDa HSP from HeLa cells, revealed the presence of an unique band, corresponding to a 74 kDa polypeptide, which was present in minimal amount in control cells and whose synthesis was increased severalfold after heat shock (Fig. 3.1d). The increase in p74 synthesis after PG treatment was comparable, and in the case of PGJ_2 superior, to that obtained after heat shock (Fig. 3.1d).

Synthesis of p74, which was minimal in control cells, was shown to increase in a dose-dependent manner in response to PGA_1 up to a concentration of 6 µg/ml, while higher concentrations apparently did not further augment it. Increase in p74 synthesis was associated with inhibition of cell proliferation but not necessarily with inhibition of protein synthesis after PGA_1 treatment (Santoro et al. 1989a). In this respect, PGA_1-mediated induction of p74 differs from HSP induction after severe heat shock, which is accompanied by a substantial inhibition of total protein synthesis (Carper et al. 1987; Black and Subjeck this Vol.).

The intracellular localization of p74 after PGA_1-treatment was investigated

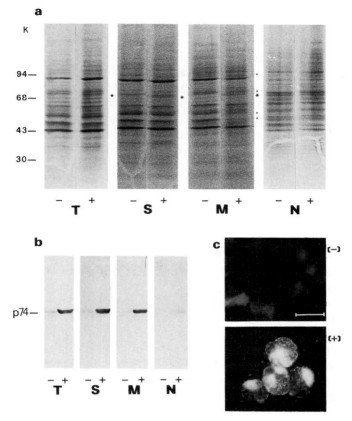

Fig. 3.2. Intracellular localization of p74 in PGA_1-treated K562 cells. Cells treated with PGA_1 (+) (4 µg/ml) or control diluent (−) for 24 h were labeled with ^{35}S-methionine, lysed, and fractionated as described in Santoro et al. (1989a). **a** Protein from total cell extracts (T), or from soluble (S), membrane (M) and nuclear (N) fractions were separated by SDS-PAGE using a 5–15% gradient gel. *Dots* indicate polypeptides that are present in reduced amount in the membrane fraction of PGA_1-treated cells. *Asterisks* indicate the p74 protein. **b** In a parallel experiment equal amounts of unlabeled proteins from total cell extracts (T) or soluble (S), membrane (M), and nuclear (N) fractions of control (−) or PGA_1-treated (+) cells were separated in SDS-PAGE and processed for immunoblot analysis using the anti-72/73kDa HSP antibody. **c** The intracellular localization of p74 is shown by indirect immunofluorescence. K562 cells treated for 24 h with diluent control (−) or PGA_1 (+) were fixed and stained with monoclonal anti-72/73kDa HSP antibody followed by fluorescein-conjugated sheep anti-mouse antibody. Methods are described in detail in Santoro et al. (1989a). (Bar = 10 µm)

by using SDS-PAGE separation of ^{35}S-methionine-labeled proteins and immunoblot analysis after cell fractionation. Figure 3.2a shows that in PGA_1-treated cells p74 accumulated in the soluble and in the membrane fractions. Moreover, although no other alteration in the pattern of the proteins synthesized in the presence of PGA_1 was found in total cell extracts or in the soluble and nuclear fractions, at least four polypeptides (97, 76, 58, and 52 kDa mol. wt., indicated by dots in Fig. 3.2a) appeared to be reduced or missing after

PGA$_1$ treatment in the membrane fraction. The presence of these proteins in the soluble, but not the membrane fraction, suggests that PGA$_1$ could interfere with the cell protein transport system. The intracellular localization of p74 in PGA$_1$-treated cells was confirmed by immunoblot analysis (Fig. 3.2b). By using this technique, a minimal amount of p74 was detected also in the nuclear fraction, and trace amounts of p74 were detected in the soluble fraction of untreated K562 cells, as reported for HSP70 in unstressed human cells throughout the cell cycle (Milarski and Morimoto 1986). The cytoplasmic localization of PGA$_1$-induced p74 was also shown by indirect immunofluorescence using the mono-clonal anti-HSP antibody (Fig. 3.2c). K562 cells grown at 37°C showed light, diffused cytoplasmic staining, as reported for asynchronous populations of HeLa cells (Milarski and Morimoto 1986). PGA$_1$ treatment caused an overall increase in the intensity of staining due to the induced p74 synthesis; p74 appeared to be localized generally in the cytoplasm and partially in small granules in the nucleus of PGA$_1$-treated cells (Fig. 3.2c). Also after PGJ$_2$ treatment p74 appears to be localized mostly in the cytoplasm, with some accumulation in dense areas of the nucleus (Santoro et al. 1989a). PG-induced p74, then, does not appear to be relocated into the nucleus, and in particular in the nucleolus, as it has been shown in cells undergoing the S-phase of the cell cycle (Milarski and Morimoto 1986), or following heat shock (Welch and Suhan 1985).

The synthesis of a heat shock-like protein of 70 kDa mol. wt. has been previously reported in K562 cells during erythroid differentiation after hemin treatment (Singh and Yu, 1984), suggesting the possibility that the p74 protein could be a marker of differentiation. Theodorakis et al. (1989) also showed that hemin induces the transcription of two members of the hsp70 gene family, hsp70 and grp78, and, by identifying the promoter elements required for hemin inducibility and the transcription factors induced during hemin stimulation, demonstrated that HSP70 induction during K562 cell maturation is through a stress response. Even though PGA$_1$ has been shown to induce erythroid differentiation in Friend erythroleukemia cells (Santoro et al. 1980a), we were unable to show the induction of erythroid differentiation of K562 cells after PGA$_1$ treatment (Santoro et al. 1986), indicating that the PGA$_1$-induced accumulation of p74 is not associated with erythroid maturation in these cells.

The data described in this section suggest instead that in human erythro-leukemia cells PG-induced p74 synthesis is associated with inhibition of cell proliferation.

Several studies focused on a possible role of HSP70 in the control of cell proliferation, but contradictory results have been described. Craig and Jacobsen (1985) reported that mutations in cognate genes of HSP70 result in a reduced growth rate in *Saccharomyces cerevisiae*. Two oncogene products, c-myc protein (Kingston et al. 1984) and adenovirus E1A 13S gene product (Nevins 1982; Wu et al. 1986) were shown to specifically induce hsp70 gene expression. HSP70 mRNA and protein synthesis are also induced by serum stimulation in quiescient HeLa cells (Wu and Morimoto 1985; Wu et al. 1987), and the hsp70 gene is transcriptionally regulated during the cell cycle in HeLa cells and in the transformed human embryonic kidney 293 cells (Kao et al. 1985; Milarski and

Morimoto 1986); in HeLa cells HSP70 is diffusely distributed in the nucleus and the cytoplasm during most phases of the cell cycle but is localized mainly in the nucleus during the S phase, suggesting a role of HSP70 in the nucleus of replicating cells (Milarski and Morimoto 1986). Moreover, in interleukin-2 (IL-2)-dependent human cells, synchronized by IL-2 deprivation, synthesis of HSP70 mRNA is increased as much as 15-fold upon interleukin-2 treatment, indicating a potential role for HSP70 in IL-2-induced cell cycle progression (Ferris et al. 1988).

In contrast, HSP70, as well as other high molecular weight HSPs, is synthesized in elevated amounts specifically in quiescent (Go) yeast cells, chicken fibroblasts, and mouse T lymphocytes, suggesting that these proteins might be involved in a cellular machinery that directs cells toward Go (Iida and Yahara 1984a, b). Similarly, Kaczmarek et al. (1987) reported that HSP70 RNA levels were elevated in quiescent peripheral human blood mononuclear cells and declined to low levels when these cells were stimulated to proliferate by serum or mitogens. Recently, Ohno et al. (1988) showed that the induction of several proteins, which were indistinguishable from heat shock-induced 68 kDa proteins by limited proteolysis, was associated with the prostaglandin-induced G_1 block of cell cycle progression of HeLa S3 cells (Ohno et al. 1988). Finally, we have recently shown that type A and J prostaglandins induce the synthesis of HSP70 also in an established cord-blood derived HTLV-I (human T-cell leukemia/lymphoma virus type I) -positive cell line, MT-2 (D'Onofrio et al. 1990b), at doses which suppress cell proliferation (D'Onofrio et al. 1990a). One of the most reasonable explanations for the discrepancies in these studies is that products of different genes of the HSP70 gene family may have different roles in the regulation of cell proliferation and cell cycle progression. The HSP70 induced by cyclopentenone prostaglandins appears to be associated with an antiproliferative state. However, there is no evidence that p74 synthesis either is responsible for the negative regulation of cell replication or is the result of the proliferation block in these systems. Although p74 synthesis can be induced by prostaglandins at concentrations which do not affect cell viability or macromolecular synthesis, the possibility cannot be excluded that it could represent the cell response to a stress situation induced by PG-treatment.

Heat shock protein synthesis by PGs could also be linked to alterations in the glycosylation process, as will be discussed in more detail in the next section. In fact we have reported that PGA_1 treatment decreased the incorporation of ^{3}H-glucosamine into K562 glycoproteins (Santoro et al. 1986), and have recently found that PGA_1 caused a more severe inhibition of ^{3}H-galactose, but not of ^{3}H-mannose, incorporation into proteins (data not shown); these changes were accompanied by alterations in protein intracellular transport, as shown by the lack of incorporation of several ^{35}S-labeled-polypeptides (Fig. 3.2a) and at least two ^{3}H-glucosamine-labeled proteins (data not shown) into K562 cell membranes. Moreover, we have recently shown that a different J-type prostaglandin, Δ^{12}-PGJ_2, which induces p74 and two other stress proteins (22 and 110 kDa mol. wt.), produced dramatic alterations in the glycosylation of K562 proteins, inhibiting the incorporation of ^{3}H-glucosamine and more severely ^{3}H-galactose in specific glycoproteins and inducing an hypermannosilation of at least three major polypeptides (210, 60, and 45 kDa mol. wt.) in these cells, suggesting an action on glycoprotein processing/maturation (Amici et al. in prep.).

HSPs and Prostaglandins in Virus Replication

A growing body of literature describes the relationship between heat shock protein synthesis and virus replication.

Induction of heat shock proteins during virus infection has been reported both in prokaryotic and eukaryotic cells. Bacteriophage infection of *E. coli*, in particular λ-phage, activates HSPs genes, and dnaK protein, the product of *E. coli* dnaK gene with a 50% homology to the *Drosophila* HSP70, is found in a complex with the phage proteins O and P, which are part of the λ-phage DNA replication complex; two other *E. coli* HSPs, the products of the groES and groEL genes, have been shown to be involved in the assembly of the head proteins of bacteriophages lambda and T4, and of the tail proteins of bacteriophage T5 (reviewed by Bond and Schlesinger 1987).

During the lytic infection of monkey and mouse cells with SV40 or polyoma virus, there is a marked increase in the synthesis of two host, heat-inducible, proteins of 92 and 72 kDa mol. wt. (Khandjian and Turler 1983). Infection of human cells with adenovirus has also been found to increase the expression of hsp genes, particularly of hsp70 genes (Kao and Nevins 1983). Mutants of *Herpes simplex* virus (HSV) type 1 (HSV1) and 2 (HSV2) also induce HSPs during infection of chicken embryo fibroblasts (Notarianni and Preston 1982) and in human neuroblastoma cells (Yura et al. 1987) respectively. The presence of abnormal forms of the HSV1 immediate early polypeptide Vmw175 was found to be the signal for induction of the stress response in chick embryo fibroblast cells (Russel et al. 1987).

Among RNA viruses, Newcastle disease virus was found to induce heat shock proteins in infected chicken embryo cells (Collins and Hightower 1982). Avirulent strains were stronger inducers and also stimulated the synthesis of glucose-regulated proteins. Peluso et al. (1977, 1978) also reported that infection of cultured chicken embryo cells by Sendai virus and simian virus 5 stimulates the synthesis of several cellular polypeptides, among which a 86 kDa protein and two glucose-regulated proteins of 99 and 78 kDa mol. wt. respectively. Induction of stress proteins by Sindbis virus and VSV in chick embryo cells was also reported by Garry et al. (1983). In this case, it was shown that the capsid protein (C) of Sindbis virus and nucleocapsid protein (N) of VSV were physically associated with a 89 kDa mol. wt. HSP. In the models that we have analyzed and that will be described later, however, we did not find induction of HSPs, and in particular of HSP70, after virus infection.

The role of HSPs in virus replication remains to be established. Induction of HSPs by viruses may simply reflect a stress from infection or may be involved in the control of virus replication.

The evidence for a relationship between prostaglandins and virus replication has been accumulating over the past few years. Viral transformation has been shown to affect prostaglandin biosynthesis in cultured cells, either increasing or decreasing it, depending on the cell type. In fact transformation of Balb/c-3T3 fibroblasts by polyoma virus (Roos et al. 1980) or by simian virus 40 (SV40) (Ritzi and Stylos 1976) was found to substantially augment PG biosynthesis, while transformation by Harvey murine sarcoma virus, Rous sarcoma virus or by SV40, greatly diminished PGE_2 production in dog kidney cells (MDCK), rat kidney cells (NRK) or rat thyroid cells (FRTL) respectively (Lin et al. 1986). The molecular events responsible for the effect of viral transformation on PG production are not known. However, it has been suggested that increased formation of prostaglandins in polyoma virus-transformed 3T3 cells is due to continuously elevated activity of phospholipase A_2 or another acyl hydrolase (Roos et al. 1980). Yaron et al. (1981) have also shown that both RNA viruses (rubella, measles, Newcastle disease) and DNA viruses (adeno) stimulate PGE production by human synovial fibroblasts, and suggested that PGE could be a mediator of virus-induced inflammation.

On the other hand, prostaglandins have been shown to influence virus replication in cultured cells, but different types of prostaglandins produce different effects on virus replication in several virus-host systems and, as shown for other aspects of prostaglandin action, their effect varies in relation to their

Table 3.1. Inhibition of RNA and DNA virus replication and induction of HSP70 synthesis by cyclopentenone prostaglandins in cultured mammalian cells

	Virus	Host cell		Reference	Induction of HSP70[a]
RNA viruses					
Paramyxoviruses	Sendai	37RC	(monkey)	Santoro et al. (1987)	+
	"	HEp-2	(human)	Santoro et al. (1983b)	ND
	"	VERO	(monkey)	Santoro et al. (1983b)	ND
Orthomyxoviruses	Influenza A (PR8)	MDCK	(canine)	Palamara et al. (1989)	+
Picornaviruses	EMC	L cells	(mouse)	Ankel et al. (1985)	ND
	Poliovirus	HeLa	(human)	Conti et al.[b]	+
	Rhinovirus	HeLa	(human)	Conti et al.[b]	+
Rhabdoviruses	VSV	L cells	(mouse)	Santoro et al. (1983a)	ND
	VSV	MA104	(monkey)	Alicea et al.[b]	+
Togaviruses	Sindbis	VERO	(monkey)	Mastromarino et al.[b]	+
	Rubella	VERO	(monkey)	Mastromarino et al.[b]	+
Retroviruses	HTLV-I[c]	CBL[d]	(human)	D'Onofrio et al. (1989)	ND
DNA viruses					
Poxviruses	Vaccinia	L cells (mouse)		Santoro et al. (1982a)	ND
Herpesviruses	HSV-I	AGMK (monkey)		Santoro (1987)	+
	HSV-II	AGMK (monkey)		Amici et al. (1988)	+
	HSV-II	HEF	(human)	Yamamoto et al. (1989)	+

[a] PGs were tested at the concentrations which inhibited virus replication.
[b] Manuscript in preparation.
[c] Human T cell leukemia/lymphoma virus-type I.
[d] Human cord blood mononuclear cells.
ND = not determined.

dose and to the structure of the cyclopentane ring (but not the number of double bonds in the aliphatic side chains), as well as the types of virus and host cells (reviewed by Santoro 1987).

As mentioned in the Introduction, in 1980 we reported that prostaglandins of the A type potently inhibit the replication of a paramyxovirus (Sendai virus) and can prevent the establishment of a persistent infection by this virus in an African green monkey kidney (AGMK) cell line, 37RC (Santoro et al. 1980b). The antiviral activity of prostaglandins of the A type and, more recently, of PGJs (Santoro et al. 1987) has now been described for several RNA and DNA viruses including orthomyxoviruses, picornaviruses, rhabdoviruses, togaviruses, pox-viruses, herpesviruses and retroviruses growing in different types of cells (Table 3.1). Recently, we have also shown that a long-acting synthetic analog of PGA_2 (16, 16-dimethyl-PGA_2 methyl ester) and PGJ_2 can suppress influenza A virus replication in mice (Santoro et al. 1988; Palamara et al. 1989). The antiviral action was specific for PGs of the A and J type, while PGs of the B, E, F series, prostacyclin, 6-keto $PGF_{1\alpha}$, and thromboxane B_2 were inactive; it was dose-dependent and decreased virus production by more than 90% at nontoxic doses which did not cause significant changes in the uptake of precursors or in the synthesis of DNA, RNA, and proteins in uninfected cells. In all virus-cell systems studied PG antiviral activity was associated with specific alterations in the synthesis and/or maturation of specific virus proteins (Santoro 1987). In mouse L fibroblasts, PGA treatment prevented the synthesis of three specific vaccinia virus (VV) polypeptides (Santoro et al. 1982a). When cytoplasmic RNAs from PGA-treated VV-infected cells were translated in cell-free systems, similar selective inhibition of viral polypeptides was observed, but PGA, even at much higher doses, did not exert any direct inhibitory action on transcription in vitro, as measured in two cell-free systems, and had no effect on primary transcription-translation of VV RNAs when assayed in a coupled cell-free systems, suggesting that the synthesis and/or activation of a host product was mediating the antiviral action (Benavente et al. 1984). PGA treatment also specifically suppressed the synthesis of the VSV glycoprotein G and altered its mobility in SDS-polyacrylamide gels, producing an apparent decrease in its molecular weight of about 4 kDa, suggesting a possible action of PGA on virus protein glycosylation (Santoro et al. 1983a).

Sendai virus, like other paramyxoviruses, consists of an inner nucleocapsid surrounded by a membrane envelope covered with spikes formed by two viral glycoproteins, the HN with both hemagglutinating and neuraminidase activity, and the F protein, which plays an essential role in hemolysis, cell fusion, and infectivity of the virion (Choppin and Scheid 1980; Peluso et al. 1978). PGA_1 and PGA_2 were able to almost totally suppress the replication of this virus in 37RC cells, at concentrations which did not inhibit protein synthesis (Fig. 3.3c) and induced the synthesis of a 74 kDa polypeptide, p74, in both Sendai virus-infected and uninfected cells (Fig. 3.3a, b) as soon as 2 h after PG treatment (Santoro et al. 1982b). PGA did not act on an early stage (i.e., adsorption, entry or uncoating) of Sendai virus replication, or by inhibiting viral (actinomycin D-resistant) RNA synthesis (Santoro et al. 1980b, 1981). SDS-PAGE analysis

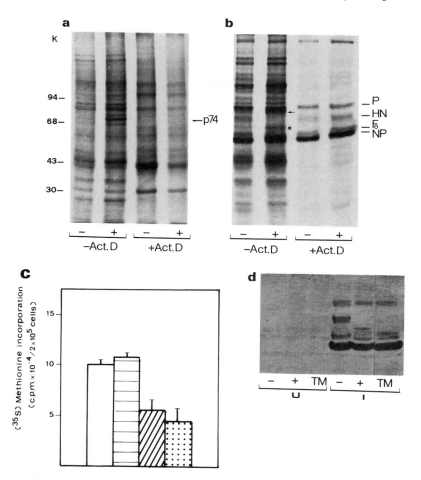

Fig. 3.3. Effect of PGA$_1$ on the synthesis of cellular and viral proteins in 37RC cells infected with Sendai virus. **a, b)** SDS-PAGE analysis of ^{35}S-methionine-labeled polypeptides of uninfected (**a**) or Sendai virus-infected (**b**) 37RC cells treated with diluent control (−) or PGA$_1$ (+) (4 µg/ml) in the presence (+ Act. D) or the absence (− Act. D) of 3 µg/ml actinomycin D. *Arrows* indicate the position of p74. HN and F$_o$ viral glycoproteins were not detected after PGA$_1$ treatment. Two new bands (66K indicated by the *asterisk*, and 63K) were instead found. The effect of PGA$_1$ on viral glycoproteins and p74 synthesis was suppressed by actinomycin D. Methods are described in detail in Santoro et al. (1989b). **c** PGA$_1$, at the dose which induced p74 synthesis, did not inhibit cell protein synthesis in uninfected or Sendai virus infected-cells. Protein synthesis was measured as incorporation into TCA-insoluble material after labeling with ^{35}S-methionine (10 µCi/10^5 cells for 24 h) 24 h after infection (p.i.). Control = □; PGA$_1$-treated = ▤; infected control = ▨; infected PGA$_1$-treated = ▥; **d** Immunoblot analysis of proteins from uninfected (*U*) or Sendai virus-infected (*I*) cells treated with control diluent (−), PGA$_1$ (+) or tunicamycin (*TM*), using a rabbit polyclonal anti-Sendai virus antibody revealed alterations in viral proteins, as described in **b**

of ^{35}S-methionine-labeled proteins showed that in PGA-treated infected cells, the three virus polypeptides P, NP, and M were synthesized normally, whereas the bands corresponding to the HN (70 kDa) and Fo (64 kDa) glycoproteins

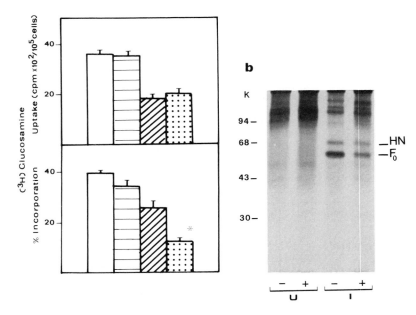

Fig. 3.4. Effect of PGA$_1$-treatment on glycoprotein synthesis in 37RC cells infected with Sendai virus. **a** Uptake by 37RC cells and incorporation into acid-insoluble material of ^{3}H-glucosamine 24 h p.i. Cells were labeled with ^{3}H-glucosamine (5 μCi/10^5 cells) for 3 h, as described in Santoro et al. (1989b). Control = □; PGA$_1$-treated = ⊟; infected control = ▨; infected PGA$_1$-treated = ⊡. **b** SDS-PAGE analysis of ^{3}H-glucosamine-labeled glycoproteins from uninfected (*U*) or Sendai virus-infected (*I*) cells treated with control diluent (−) or PGA$_1$ (+) (4 μg/ml). Equal amount of radioactivity for each sample was applied to the gels to be able to detect virus proteins in PGA$_1$-treated cells. Taking into consideration the amounts of protein applied to the gels, about 10–20% of glycosylated HN and F$_o$ proteins are detected in PGA$_1$-treated cells. (Santoro et al. 1989b)

were absent and two new viral proteins of molecular weight 66 and 63 kDa respectively were detected (Fig. 3.3a, b). These results were confirmed by immunoblot analysis using a rabbit polyclonal anti-Sendai antibody (Fig. 3.3d). These two new proteins were shown to be altered forms of the viral glyco-proteins HN and Fo, and the alterations were found to be related to a defect in the glycosylation process, as demonstrated by a dramatic inhibition of ^{3}H-glucosamine in Sendai virus-infected (but not in uninfected) cells and specifically in the two virus glycoproteins HN and Fo (Fig. 3.4a, b). This effect was not due to a decrease in the uptake of sugar by the cells and, even though PGA appeared to act on an early step of viral protein glycosylation, since no intermediate forms of glucosamine-labeled virus proteins were detected, its action was shown not to be a tunicamycin-like effect (Santoro et al. 1989b). The PGA-induced structural alteration of HN inhibits its incorporation into the host cell membrane and consequently virus assembly and budding from 37RC cells (Santoro et al. 1989b). Actinomycin D suppressed PGA-induced synthesis of p74, suggesting that transcription was necessary (Fig. 3.3a). In virus-infected cells actinomycin D

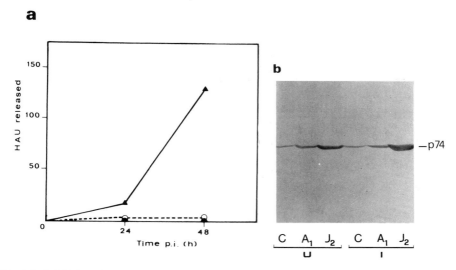

Fig. 3.5. Inhibition of Sendai virus production by PGs (**a**) and identification of p74 as a HSP70 by immunoblot analysis in 37RC cells. **a** 37RC cells were treated with PGA$_1$ (4 µg/ml) (○), PGJ$_2$ (4 µg/ml) (●) or diluent control (▲) 1 h p.i. Virus production at different times p.i. was determined by measuring the hemagglutinin units (*HAU*) in the culture medium, as previously described (Santoro et al. 1989b). **b** Equal amounts of proteins from uninfected (*U*) or Sendai virus-infected (*I*) cells treated with PGA$_1$ (A$_1$), PGJ$_2$ (J$_2$) or diluent control (*C*), were separated by SDS-PAGE and processed for immunoblot analysis using a monoclonal anti 72/73kDa HSP antibody from HeLa cells

treatment, which suppresses cellular but not viral RNA synthesis, abolished both the PGA$_1$-induced production of the p74 protein and the alterations in the synthesis of viral glycoproteins (Fig. 3.3a, b), suggesting that, as in the case of vaccinia virus in L cells, the synthesis of a host product, p74, mediated the effect of PGA$_1$ on viral proteins.

p74 has been recently identified as a heat shock protein related to the HSP70 group by immunoblot analysis using monoclonal antibodies against the human HSP72/73 from HeLa cells (Fig. 3.5b). This protein appears to be constitutively synthesized in 37RC cells at significant basal level and its synthesis is increased several fold by PGA$_1$ and even more dramatically by PGJ$_2$ treatment. The possibility that p74 could be identical to the glucose-regulated grp78 was excluded by the fact that grp78, induced by tunicamycin in these cells, is characterized by a different mobility in SDS-PAGE (unpubl. data). Even though the possibility cannot be excluded that synthesis of p74 simply represents the cellular response to a stress situation following exposure to PGs, a possible role for this protein in the antiviral activity of prostaglandins is suggested by several observations: (1) p74 synthesis is induced only by prostaglandins with antiviral activity (A and J type PGs) and moderately by the PG precursor arachidonic acid, while PGs of the B, E, F series, prostacyclin and thromboxane B$_2$, have no effect; (2) PG induction of p74 is dose-dependent and starts at doses which inhibit virus replication; (3) suppression of p74 synthesis by actinomycin D abolished the PGA-induced alterations of virus protein synthesis (Santoro et al. in prep.). However, there is no evidence that p74 has a direct role in inhibiting virus replication or is synthesized in response to the accumulation of abnormal or incorrectly glycosylated proteins after PG treatment. The presence of abnormal or defectively glycosylated proteins is, in fact, a well-known signal for induction of HSP synthesis (reviewed by Pelham 1986). Δ^{12}-PGJ$_2$, a dehydration product of PGD$_2$, which has also been recently shown to possess potent antiviral activity in several virus-host models (Table 3.1), induced p74 and at least three additional proteins whose molecular weights (36, 92, and 112 kDa respectively) were similar to other stress proteins, suggesting that specific PGs can induce one or more HSPs.

Concluding Remarks

Induction of HSP70 by cyclopentenone PGs has been shown in two monkey cell lines, 37RC and MA 104, a dog kidney cell line (MDCK) and several human cell lines among which K562 erythroleukemia, MT-2 cells, HeLa cells, and human embryonic fibroblasts (see second section and Table 3.1), suggesting that synthesis of heat shock proteins could be a general response of mammalian cells to PGs.

Even though the data suggest that HSP70 synthesis could be associated with the antiproliferative and the antiviral activity of prostaglandins, its role remains to be established. The fact that both actions of PGs are accompanied by alterations in the synthesis and/or glycosylation of specific proteins suggests that accumulation of abnormal proteins could be responsible for HSP induction. However, the possibility that prostaglandins could induce HSP synthesis by other mechanisms cannot be excluded. It has, in fact, been shown that PGs of the A and J type are transported into the nuclei where they bind to nuclear proteins (Narumiya and Fukushima 1986; Narumiya et al. 1986, 1987) and to DNA (Karmali et al. 1976). They could alter chromatin structure and/or the nuclear matrix and thereby lead to hsp gene activation.

Finally, because of their ubiquitous nature, their pleiotropic action, their increased synthesis after different kinds of stresses (among which heat shock, Chen et al. 1987), and their well-known physiological role in the production of fever (Dinarello and Wolff 1982; Coceani et al. 1989), prostaglandins could be considered as possible intracellular signals for induction of heat shock proteins.

Acknowledgements

This work was partially supported by a grant from the Italian National Council of Research, Progetto Finalizzato FATMA.

References

Amici C, Benedetto A, Garaci E, Santoro MG (1988) Prostaglandins of the A and J series inhibit herpes simplex virus II replication in vitro. In: 6th Mediterr Congr Chemotherapy, 22–27 May 1988, Taormina, Italy, (Abstr) p 215

Ankel H, Mittnacht S, Jacobsen H (1985) Antiviral activity of Prostaglandin A on encephalomyocarditis virus-infected cells: a unique effect unrelated to interferon. J Gen Virol 66:2355–2364

Benavente J, Esteban M, Jaffe BM, Santoro MG (1984) Selective inhibition of viral gene expression as the mechanism of the antiviral action of PGA_1 in vaccinia virus-infected cells. J Gen Virol 65:599–608

Bhuyan BK, Adams EG, Badiner GJ, Li LH, Barden K (1986) Cell cycle effects of prostaglandins A_1, A_2 and D_2 in human and murine melanoma cells in culture. Cancer Res 46:1688–1693

Bond U, Schlesinger MJ (1987) Heat-shock proteins and development. Adv Genet 24:1–29

Bregman MD, Funk C, Fukushima M (1986) Inhibition of human melanoma growth by prostaglandin A, D and J analogues. Cancer Res 46:2740–2744

Breitman TR (1987) The role of prostaglandins and other arachidonic acid metabolites in the differentiation of HL-60. In: Garaci E, Paoletti R, Santoro MG (eds) Prostaglandins in Cancer Research. Springer, Berlin Heidelberg New York Tokyo, pp 161–171

Burdon RH (1982) The human heat-shock proteins: their induction and possible intracellular functions. In: Schlesinger MJ, Ashburner M, Tissieres A (eds) Heat shock from bacteria to man. Cold Spring Harbor Lab, New York, pp 283–288

Carper SW, Duffy JJ, Gerner EW (1987) Heat shock proteins in thermotolerance and other cellular processes. Cancer Res 47:5249–5255

Chen BD, Sapareto SA, Chou T (1987) Induction of prostaglandins production by hyperthermia in murine peritoneal exudate macrophages. Cancer Res 47:11–15

Choppin PW, Scheid A (1980) The role of viral glycoproteins in adsorption, penetration and pathogenicity of viruses. Rev Infect Dis 2:40–61

Coceani F, Bishai I, Lees J, Sirko S (1989) Prostaglandin E_2 in the pathogenesis of pyrogen fever: validation of an intermediary role. In: Samuelsson B, Wong PYK, Sun FF (eds) Advances in prostaglandins, thromboxane and leukotriene research, Vol 19. Raven Press, New York, pp 394–397

Collins PL, Hightower L (1982) Newcastle Disease Virus stimulates the cellular accumulation of stress (heat shock) proteins. J Virol 44:703–707

Craig EA, Jacobsen K (1985) Mutations in cognate genes of *Saccharomyces cerevisiae* hsp70 result in reduced growth rates at low temperatures. Mol Cell Biol 5:3517–3524

Dinarello CA, Wolff SM (1982) Molecular basis of fever in humans. Am J Med 72:799–819

D'Onofrio C, Bonmassar E, Santoro MG (1989) Inhibition of HTLV-I infection in vitro by antiviral prostaglandins A and J. Pharmacol Res 21:665–666

D'Onofrio C, Alvino E, Garaci E, Bonmassar E, Santoro MG (1990a) Selection of HTLV-I positive clones is prevented by prostaglandin A in infected cord blood cultures. Br J Cancer 61:207–214

D'Onofrio C, Amici C, Bonmassar E, Santoro MG (1990b) The antiproliferative effect of prostaglandins A and J on HTLV-I transformed cells is associated with induction of a heat shock protein. Pharmacol Res (in press)

Ferris DK, Harel-Bellan A, Morimoto R, Welch WJ, Farrar WI (1988) Mitogens and lymphokine stimulation of heat shock proteins in T lymphocytes. Proc Natl Acad Sci USA 85:3850–3854

Fukushima M, Kato I, Narumiya S et al. (1989) Prostaglandins A and J: antitumor and antiviral prostaglandins. In: Samuelsson B, Wong PY, Sun FF (eds) Advances in prostaglandins, thromboxane and leukotriene research, Vol 19. Raven Press, New York, pp 415–418

Garaci E, Paoletti R, Santoro MG (1987) Prostaglandins in cancer research. Springer, Berlin Heidelberg New York Tokyo.

Garry RF, Emin TU, Bose HR (1983) Induction of stress proteins in Sindbis virus- and vesicular stomatitis virus-infected cells. Virology 129:319–332

Honma Y, Kasukabe T, Hozumi M, Koshihara Y (1980) Regulation of prostaglandin synthesis during differentiation of cultured mouse myeloid leukemia cells. J Cell Physiol 104:349–357

Hughes-Fulford M (1987) The mechanism of prostaglandin inhibition on the cell cycle. In: Garaci E, Paoletti R, Santoro MG (eds) Prostaglandins in cancer research. Springer, Berlin Heidelberg New York Tokyo, pp 115–128

Hughes-Fulford M, Wu J, Kato T, Fukushima M (1985) Inhibition of DNA synthesis and cell cycle by prostaglandins independent of cyclic AMP. In: Hayaishi O, Yamamoto S (eds) Advances in prostaglandins, thromboxane and leukotriene research, Vol 15. Raven Press, New York, pp 401–404

Iida H, Yahara I (1984a) Specific early-G_1 blocks accompanied with stringent response in *Saccharomyces cerevisiae* lead to growth arrest in resting state similar to the Go of higher eukaryotes. J Cell Biol 98:1185–1193

Iida H, Yahara I (1984b) Durable synthesis of high molecular weight heat shock proteins in Go cells of the yeast and other eukaryotes. J Cell Biol 99:199–207

Johnson GS, Pastan I (1971) Change in growth and morphology of fibroblast by prostaglandins. J Natl Cancer Inst 47:1357–1360

Kaczmarek L, Calabretta B, Kao H, Heintz N, Nevins J, Baserga R (1987) Control of hsp70 RNA levels in human lymphocytes. J Cell Biol 104:183–187

Kao H, Nevins JR (1983) Transcriptional activation and subsequent control of the human heat shock gene during adenovirus infection. Mol Cell Biol 3:2058–2065

Kao H, Capasso O, Heintz N, Nevins JR (1985) Cell cycle control of the human HSP70 gene: implications for the role of a cellular E1A-like function. Mol Cell Biol 5:628–633

Karmali R, Schiller P, Horrobin DF (1976) Prostaglandins can prevent the binding of chloroquine to calf thymus DNA. Prostaglandins 12:463–469

Khandjian EW, Turler H (1983) Simian virus 40 and polyoma virus induce the synthesis of heat shock proteins in permissive cells. Mol Cell Biol 3:1–8

Kingston RE, Baldin AS, Sharp PA (1984) Regulation of heat-shock protein 70 gene expression by c-myc. Nature (Lond) 312:280–282

Lin MC, Segawa K, Ito Y, Beckner SK (1986) The effect of viral transformation on prostaglandin production depends on the cell type. Virology 155:19–26

Lozzio CB, Lozzio BB (1975) Human chronic myelogenous leukemia cell-line with positive Philadelphia chromosome. Blood 45:321–334

Marini S, Palamara AT, Garaci E, Santoro MG (1990) Growth inhibition of Friend erythroleukemia cell tumours in vivo by a synthetic analogue of prostaglandin A: an action independent of natural killer activity. Br J Cancer 61:394–399

Milarski KL, Morimoto R (1986) Expression of human HSP70 during the synthetic phase of the cell cycle. Proc Natl Acad Sci USA 83:9517–9521

Moore MAS (1982) G-CSF: its relationship to leukemia differentiation-inducing activity and other hemopoietic regulators. J Cell Physiol 1:53–64

Narumiya S, Fukushima M (1985) Δ^{12}-prostaglandin J_2, an ultimate metabolite of prostaglandin D_2 exerting cell growth inhibition. Biochem Biophys Res Commun 127:739–745

Narumiya S, Fukushima M (1986) Site and mechanism of growth inhibition by prostaglandins. I. Active transport and intracellular accumulation of cyclopentenone prostaglandins, a reaction leading to growth inhibition. J Pharmacol Exp Ther 293:500–505

Narumiya S, Ohno K, Fujiwara M, Fukushima M (1986) Site and mechanism of growth inhibition by prostaglandins. II. Temperature-dependent transfer of cyclopentenone prostaglandin to nuclei. J Pharmacol Exp Ther 293:506–511

Narumiya S, Ohno K, Fukushima M, Fujiwara M (1987) Cellular uptake and nuclear accumulation of prostaglandin A and J, a mechanism of prostaglandin-induced growth inhibition. In: Garaci E, Paoletti R, Santoro MG (eds) Prostaglandins in cancer research. Springer, Berlin Heidelberg New York Tokyo, pp 86–96

Nevins JR (1982) Induction of the synthesis of a 70,000 dalton mammalian heat shock protein by the adenovirus E1A gene product. Cell 29:913–919

Ninnemann JL (1988) Prostaglandins, leukotrienes and the immune response. Cambridge Univ Press, Cambridge New York

Notarianni EL, Preston CM (1982) Activation of cellular stress protein genes by herpes simplex virus temperature-sensitive mutants which overproduce early polypeptides. Virol 123:113–122

Ohno K, Fujiwara M, Fukushima M, Narumiya S (1986) Metabolic dehydration of prostaglandin E_2 and cellular uptake of the dehydration product: correlation with prostaglandin E_2-induced growth inhibition. Biochem Biophys Res Commun 139:808–815

Ohno K, Fukushima M, Fujiwara M, Narumiya S (1988) Induction of 68,000 dalton heat shock proteins by cyclopentenone prostaglandins. J Biol Chem 263:19764–19770

Olsson IL, Breitman TR, Gallo RC (1982) Priming of human myeloid leukemia cell line HL-60 and U-937 with retinoic acid for differentiation effects of cyclic adenosine 3′:5′-monophosphate-inducing agents and a T-lymphocyte-derived differentiation factor. Cancer Res 42:3928–3933

Palamara AT, Pica F, Amici C, Figna L, Garaci E, Santoro MG (1989) In vitro and in vivo antiviral activity of delta 12-prostaglandin J_2. In: Int Symp Antiviral Chemotherapy, 1–5 October 1989, Porto Cervo, Italy, (Abstr) p 15

Pelham HR (1986) Speculations on the function of the major heat shock and glucose-regulated proteins. Cell 46:959–961

Peluso RW, Lamb RA, Choppin PW (1977) Polypeptide synthesis in simian virus 5-infected cells. J Virol 23:177–187

Peluso RW, Lamb RA, Choppin PW (1978) Infection with paramyxoviruses stimulates synthesis of cellular polypeptides that are also stimulated in cells transformed by Rous sarcoma virus or deprived of glucose. Proc Natl Acad Sci USA 75:6120–6124

Ritzi EM, Stylos WA (1976) Prostaglandin production in cultures of BALB/3T3 and SV3T3 mouse fibroblasts. JNCI 56:529–533

Robert A (1981) Prostaglandins and the gastrointestinal tract. In: Johnson LR (ed) Physiology of the gastrointestinal tract. Raven Press, New York, pp 1407–1434

Roos P, Lindgren JA, Hammarström S (1980) On the mechanism of elevated prostaglandin E_2 production in 3T3 fibroblast transformed by polioma virus. Eur J Biochem 108:279–283

Russel J, Stow EC, Stow ND, Preston CM (1987) Abnormal forms of the herpes simplex virus immediate early polypeptide Vmw 175 induce the cellular stress response. J Gen Virol 68:2397–2406

Ruwart J, Rush BD, Friedle NM, Piper RD, Kolaja GJ (1981) Protective effects of 16, 16-dimethyl PGE_2 on the liver and kidney. Prostaglandins 21:97–102

Samuelsson B (1982) Prostaglandins, thromboxanes and leukotrienes: biochemical pathways. In: Powles TJ, Bochman RS, Honn KV, Ramwell P (eds) Prostaglandins and cancer: First International Conference. Alan R Liss Inc, New York, pp 1–19

Santoro MG (1987) Involvement of protein synthesis in the antiproliferative and the antiviral action of prostaglandins. In: Garaci E, Paoletti R, Santoro MG (eds) Prostaglandins in cancer research. Springer, Berlin Heidelberg New York Tokyo, pp 97–114

Santoro MG, Philpott GW, Jaffe BM (1976) Inhibition of tumor growth in vivo and in vitro by prostaglandin E. Nature (Lond) 263:777–779

Santoro MG, Benedetto A, Jaffe BM (1980a) The role of prostaglandins in Friend erythroleukemia cell growth and differentiation. In: Rossi GB (ed) Erythropoiesis and differentiation in Friend leukemia cells. Elsevier/North Holland Press, Amsterdam New York, pp 553–562

Santoro MG, Benedetto A, Carruba G, Garaci E, Jaffe BM (1980b) Prostaglandin A compounds as antiviral agents. Science 209:1032–1034

Santoro MG, Carruba G, Garaci E, Jaffe BM, Benedetto A (1981) Prostaglandins of the A series inhibit Sendai virus replication in cultured cells. J Gen Virol 53:75–83

Santoro MG, Jaffe BM, Garaci E, Esteban M (1982a) Antiviral effect of prostaglandins of A series: inhibition of vaccinia virus replication in cultured cells. J Gen Virol 63:435–440

Santoro MG, Jaffe BM, Elia G, Benedetto A (1982b) Prostaglandin A_1 induces the synthesis of a new protein in cultured AGMK cells. Biochem Biophys Res Comm 107:1179–1184

Santoro MG, Jaffe BM, Esteban M (1983a) Prostaglandin A inhibits the replication of vesicular stomatitis virus: effect on virus glycoprotein. J Gen Virol 64:2797–2801

Santoro MG, Benedetto A, Zaniratti S, Garaci E, Jaffe BM (1983b) The relationship between prostaglandins and virus replication. Prostaglandins 25:353–364

Santoro MG, Crisari A, Benedetto A, Amici C (1986) Modulation of the growth of a human erythroleukemic cell line (K562) by prostaglandins: antiproliferative action of PGAs. Cancer Res 46:6073–6077

Santoro MG, Fukushima M, Benedetto A, Amici C (1987) PGJ_2, a new antiviral prostaglandin: inhibition of Sendai virus replication and alteration of virus protein synthesis. J Gen Virol 68:1153–1158

Santoro MG, Favalli C, Mastino A, Jaffe BM, Esteban M, Garaci E (1988) Antiviral activity of a synthetic analog of prostaglandin A in mice infected with influenza A virus. Arch Virol 99:89–100

Santoro MG, Garaci E, Amici C (1989a) Prostaglandins with antiproliferative activity induce the synthesis of a heat shock protein in human cells. Proc Natl Acad Sci USA 86:8407–8411

Santoro MG, Amici C, Elia G, Benedetto A, Garaci E (1989b) Inhibition of virus protein glycosylation as the mechanism of the antiviral action of prostaglandin A_1 in Sendai virus-infected cells. J Gen Virol 70:789–800

Shahabi NA, Chegini N, Wittliff JL (1987) Alterations of MCF-7 human breast cancer after prostaglandins PGA_1 and $PGF_{2\alpha}$ treatment. Exp Cell Biol 55:18–27

Singh MK, Yu J (1984) Accumulation of a heat shock-like protein during differentiation of human erythroid cell line K562. Nature (Lond) 309:631–633

Theodorakis NG, Zand DJ, Kotzbauer PT, Williams GT, Morimoto RI (1989) Hemin-induced transcriptional activation of the HSP70 gene during erythroid maturation in K562 cells is due to a heat shock factor-mediated stress response. Mol Cell Biol 9:3166–3173

Vane JR (1987) Antiinflammatory drugs and the arachidonic acid cascade. In: Garaci E, Paoletti R, Santoro MG (eds) Prostaglandins in cancer research. Springer, Berlin Heidelberg New York Tokyo, pp 12–28

Welch WJ, Suhan JP (1985) Morphological studies of the mammalian stress response: characterization of changes in cytoplasmic organelles, cytoskeleton and nucleoli and appearance of intranuclear active filaments in rat fibroblast after heat shock. J Cell Biol 101:1198–1211

Wu BJ, Morimoto RI (1985) Transcription of the human hsp70 gene is induced by serum stimulation. Proc Natl Acad Sci USA 82:6070–6074

Wu BJ, Hurst HC, Jones NC, Morimoto RI (1986) The E1A 13S product of adenovirus-5 activates transcription of the cellular human HSP70 gene. Mol Cell Biol 6:2994–2999

Wu BJ, Williams G, Morimoto RI (1987) Detection of three protein binding sites in the serum-regulated promoter of the human gene encoding the 70kDa heat-shock protein. Proc Natl Acad Sci USA 84:2203–2207

Yamamoto N, Rahman M, Fukushima M, Maeno K, Nishiyama Y (1989) Involvement of prostaglandin-induced proteins in the inhibition of herpes simplex virus replication. Biochem Biophys Res Commun 158:189–194

Yaron M, Yaron I, Caspi D, Smentana O, Eylan E, Zor U (1981) RNA and DNA viral stimulation of prostaglandin E production by human synovial fibroblasts. Arthr Rheuma 24:1582–1586

Yura Y, Terashima K, Iga H, Kondo Y, Yanagawa T, Yoshida H, Hiyashi Y, Sato M (1987) Macromolecular synthesis at the early stage of herpes simplex type 2 (HSV 2) latency in a human neuroblastoma cell line IMR-32: repression of late viral polypeptide synthesis and accumulation of cellular heat-shock proteins. Arch Virol 96:17–28

CHAPTER 4

Heat Shock and Adaptation During Temperature-Activated Dimorphism in the Fungus *Histoplasma capsulatum*

Bruno Maresca and Luisella Carratù,

International Institute of Genetics and Biophysics, Via Marconi, 12, 80125 Naples, Italy

Histoplasma capsulatum is the causative agent of histoplasmosis, a systemic fungal disease worldwide in occurrence, and the most common respiratory mycotic infection affecting humans and animals (Schwarz 1981). Several fungi, in particular those pathogenic such as *H.capsulatum*, *Blastomyces dermatitidis*, *Paracoccidioides brasiliensis* etc., can assume a filamentous or unicellular morphology in response to changes in the environmental conditions (Maresca and Kobayashi 1989). *H.capsulatum* grows as mycelia in soil while the yeast phase is the only form present in patients. In laboratory conditions, dimorphism is directly and reversibly controlled by temperature changes. The temperature-induced transition and the events in establishment of infection seem to be intimately correlated. In fact, the temperature works not only as a signal for adaptation through the induction of a heat shock-like phenomenon, but also in triggering the phase transition. The role that the heat shock response plays during the differentiation process and in the adaptation to high temperature in *H.capsulatum* will be discussed here.

Introduction

The heat shock phenomenon is a physiological and biochemical response of all prokaryotic and eukaryotic cells to an abrupt increase in temperature (Ashburner and Bonner 1979; Lindquist 1986) and to a variety of changes in other environmental conditions (Craig 1985). The response is generally characterized by a repression of normal protein synthesis and by a very rapid initiation of transcription of several heat shock protein-encoding genes (Lindquist 1986). The protein sequences and the mode of regulation of these stress-induced proteins have been highly conserved throughout evolution. Furthermore, sudden increase in the environmental temperature has also important consequences in modulating the activity of a number of membrane-associated functions, like membrane shape and composition, redox potential, and mitochondrial electron transport (Alberts et al. 1983; Patriarca and Maresca 1990), which play a particularly important role in dimorphic organisms such as *H.capsulatum*.

One of the most interesting aspects of the heat shock response is the manner

Stress Proteins
Schlesinger, Santoro, Garaci (Eds.)
© Springer-Verlag Berlin Heidelberg 1990

in which temperature intervenes to control gene expression and to induce the homeostatic effect. In most of the eukaryotic systems the temperature must be lowered to physiological levels after heat shock to ensure survival. Therefore the homeostatic effect induced by the heat shock proteins is temporary in most biological systems (Ashburner and Bonner 1979; Lindquist 1986). On the other hand in parasitic organisms, following the host invasion, the adaptation to the different biochemical conditions, including a higher environmental temperature, is a necessary step to allow phase transition and to establish the pathogenic form. Thus, in these organisms the temperature plays a crucial role in controlling not only the heat shock response but also the process of adaptation to the new environment during the differentiation process.

Temperature Adaptation in *H.capsulatum*

H.capsulatum is the most extensively studied of the dimorphic pathogenic fungi, with a parasitic phase consisting of yeast cells and a saprobic mycelial phase. The morphological transition can be reversibly accomplished by shifting the temperature of incubation from 25 °C (mycelia cells) to 37 °C (yeast cells). However, an increase in temperature is not sufficient by itself to ensure the establishment of the yeast phase, e.g., components such as −SH groups must be present in the growth medium (Maresca and Kobayashi 1989). In particular, the temperature seems to affect the redox state of −SH groups or the general redox potential of cells, or both, which by controlling specific molecular mechanisms would determine the phase of the organism. In fact, while the induction of the mycelial to yeast phase can occur in the absence of reduced −SH-containing groups and molecular oxygen, cysteine and oxygen are required later to complete the transition and support yeast growth (McVeigh and Houston 1972). It is possible to speculate that the dependence of the yeast phase on cysteine may reflect an adaptation to a very specific environment. Yeast cells are found in macrophages in which the intracellular environment has high reducing capacity as compared with that encountered by mycelia, which live in soil rich in oxidized compounds. Therefore, the requirement for cysteine provides a biological control point which would ensure that an adequate level of this amino acid is present for transition to proceed (Maresca et al. 1981). Available data, in fact, suggest that specific "check point(s)" exist(s) in *H.capsulatum* during the mycelium to yeast transition which allow the organism to determine whether or not it is really in the new environment: for example, a temperature increase should not simply be due to fluctuations in the external environment (night/day, exposure to shade or sun) but determined by host invasion. Therefore, a change in temperature should not be "read" by the fungus as the necessary and sufficient condition to start the differentiation process. Temperature would initiate a series of reactions that permits the organism to proceed towards a new cell stage only if other conditions, like the presence or absence of growth factors, redox potential etc., are satisfied.

Further, yeast cultures of *H.capsulatum* spontaneously transform to mycelia when they reach the stationary state at 37 °C. In the same growth medium, an over-grown culture of mycelia incubated at 25 °C does not transform spontaneously to the yeast phase at this temperature. Therefore, mycelia to yeast transition and the maintenance of the yeast phase require a well-defined temperature. This supports the idea that "mycelia-specific genes" may be expressed at any temperature, while "yeast and transition-specific genes" are regulated and can be expressed only at high temperature in the presence of cysteine in the growth medium (Maresca and Kobayashi 1989).

Dimorphism in fungi is a complex mechanism that includes, in addition to the changes in the regulation of genes directly involved in morphogenesis, modifications in those coding for functions necessary for the adaptation to the new living conditions and for the induction of the pathogenicity. As will be described later, at least two important sets of modifications occur during the shift from mycelia to yeast cells: profound changes in the mitochondrial activities and the induction of the heat shock phenomenon. It appears that these molecular changes represent "unit circuits", each one dealing primarily with a single phenomenon: the effect of temperature on mitochondrial electron transport efficiency and energy synthesis, and the induction of heat shock genes. Progress toward the late stages of morphogenesis would then be contingent on these circuits, but not result from them. These physiological genetic compartments may not themselves be instructive in the regulation of specific patterns of gene expression necessary for mycelial to yeast phase transition, but they may be critical for the adaptation of the organism to the new environmental temperature at which new patterns of gene expression take hold (Maresca and Kobayashi 1989). Further, the induction of the thermotolerant state (induced either by mild temperatures or starvation) influences the mitochondrial membrane functions as demonstrated by the ability to protect ATPase activity during severe heat shock conditions (Patriarca and Maresca 1990).

Heat Shock and Electron Transport Efficiency

During the first few hours of the increase in temperature of incubation in *H.capsulatum*, cell metabolism slows drastically and cells eventually enter a dormant state as defined by the absence of measurable cytochrome components, RNA and protein synthesis, and oxygen consumption (Maresca et al. 1981). However, even though the morphological mycelium to yeast phase transition occurs within 6 to 8 days according to the strain, critical biochemical changes take place immediately after the temperature shift: among others, oxidative phosphorylation becomes uncoupled and a rapid decline in the intracellular ATP concentration occurs. We and other authors have found that this phenomenon occurs in all of the strains so far analyzed (Medoff et al. 1986). Further, the temperature at which the cells enter the dormant stage is specific for a given strain and is dependent on its temperature sensitivity and this in turn correlates

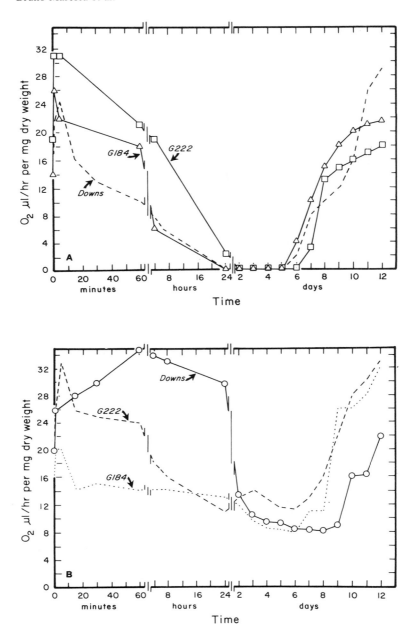

Fig. 4.1. Oxygen consumption at various time points during the mycelial to yeast phase transition. The experiment was initiated by shifting early log phase cultures of mycelia from 25 to 34 and 37 °C (Downs strain) or to 37 and 40 °C (G222B and G184A strains). Respiration was measured polarographically using a KIC oxygraph equipped with a Clark-type oxygen electrode (Gilson). Cells were resuspended in respiration buffer containing 1% mannose, 1 mM CaCl$_2$, 1 mM dimethylglutaric acid (pH 7.2). Cell respiration rate was expressed as μl O$_2$/h per mg dry weight. **A** shows respiration rates of Downs strain after the temperature shift from 25 to 37 °C compared with those of G222B and G184A strains shifted from 25 to 40 °C. **B** shows Downs when shifted from 25 to 34 °C and G222B and G184A from 25 to 37 °C. (Medoff et al. 1986)

to its virulence. In fact, it has been observed that the higher the temperature required to turn off cell metabolism in a given strain, the higher the level of its pathogenicity (Medoff et al. 1986). For example, the less virulent Downs strain (LD_{50} for AKR mice 7 days after infection is $20-22 \times 10^6$ yeast ml) undergoes extreme metabolic changes when the temperature is shifted to 37 °C, while a higher temperature is required for virulent strains (Fig. 4.1a). In fact, in G184A, G186A, G217B, and G222B (with an LD_{50} for AKR mice between 0.5 and 9.0×10^6 yeast/ml) a temperature between 39 and 43 °C is required to observe a similar phenomenon (Lambowitz et al. 1983; Medoff et al. 1986). Therefore, the physiological changes of more virulent strains at 37 °C are fundamentally similar but less extreme than those in the Downs strain and seem to be dependent on the capacity of the cells to induce an "appropriate" heat shock response. In fact, the behavior of the Downs strain can be made to resemble that of the virulent strains at 37 °C simply by shifting it to the lower temperature of 34 °C (Medoff et al. 1986; Fig. 4.1b). At these conditions, cells recover to yeast phase in a shorter period of time than that measurable at extreme temperatures. As we will see later, the capacity of inducing heat shock response at an intermediate temperature (induction of thermotolerance), detected by hybridization of RNA extracted at 34 and 37 °C with hsp70 and hsp83 genes, show a strong correlation between the temperature sensitivity (and virulence), the extent of the biochemical and physiological changes and the level of heat shock genes transcription at a specific temperature (Patriarca et al. 1987; Patriarca and Maresca 1990).

Since the mycelial to yeast transition is induced by a sudden change in temperature to 37 °C, Lambowitz et al. suggested that the early event of this transition in *H. capsulatum* was a "heat shock response" which was followed by cell adaptation to the higher temperature (Lambowitz et al. 1983). They postulated that the triggering event for the biochemical and morphological changes that take place immediately after the temperature shift was the rapid decline in intracellular ATP levels that follows the uncoupling of oxidative phosphorylation. These results led to the new concept that morphogenesis in general (or at least the regulation of adaptation to the human body temperature of 37 °C) in pathogenic fungi and in other diphasic organisms may simply be a by-product of the heat shock response (Lambowitz et al. 1983). These authors have proposed that heat shock proteins may contribute, among others, to the adaptation to the new temperature in the host in the early hours of morphogenesis. We and other authors have done a more detailed study on coupling of oxidative phosphorylation on other more virulent strains (Medoff et al. 1986). As demonstrated for the efficiency of respiration during phase transition, we found that capacity of maintaining coupled oxidative phosphorylation to electron transport was dependent on the temperature and on the strain used. In fact, Fig. 4.2 shows that, contrary to what happens to the Downs strain, 1 min after the shift from 25 to 37 °C, in the more pathogenic G184A strain respiration remains tightly coupled (Fig. 4.2a) and a temperature of 41 °C is necessary to uncouple ATP synthesis (Fig. 4.2b). Since the conversion to the yeast phase appears to be required for infection, it would be likely that the rapidity of the transformation to the yeast phase would also be important for virulence.

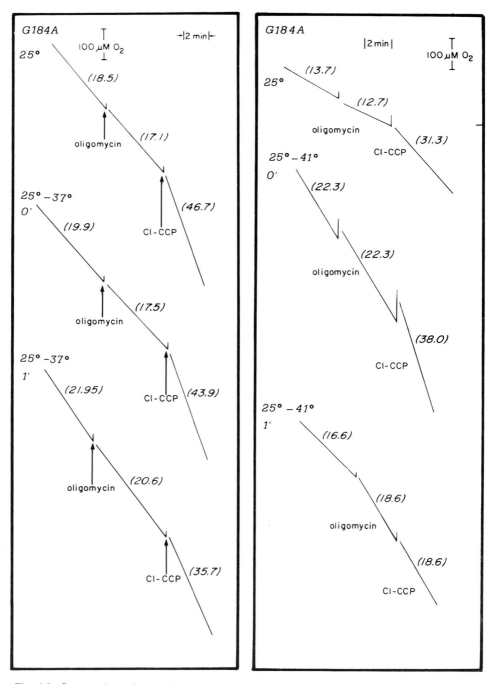

Fig. 4.2. Oxygen electrode recordings of respiration of mycelial cells at 25 °C, immediately after (0) and 1 min after shift up to 34 and 37 °C (Downs) or to 37 and 41 °C (G184A). Coupling of oxidative phosphorylation was assayed by the ability of the uncoupler carbonyl cyanide *m*-chorophenyl-hydrazone (Cl-CCP, Sigma) to stimulate respiration in the presence of oligomycin (Sigma), an inhibitor of ATP synthetase. Cell respiration assays were carried out in the presence of salicyl-hydroxamic acid (SHAM) to inhibit the alternate oxidase and force electron flux through the cyto-

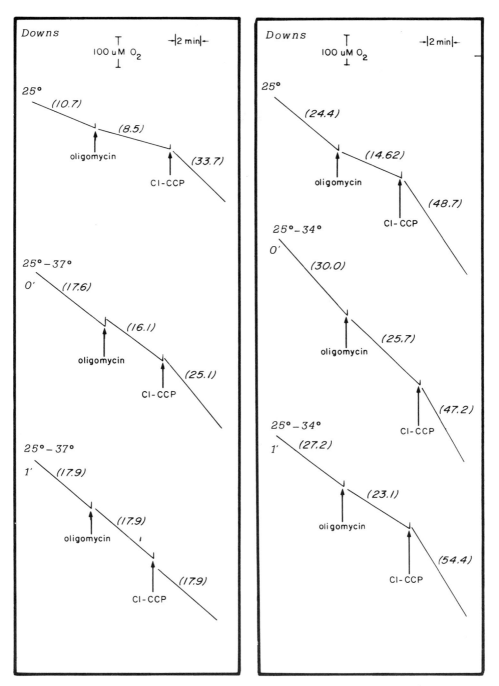

chrome system (Maresca et al. 1979). SHAM was added at 0.2 mM, oligomycin at 5 µg/ml and Cl-CCP at 0.5 mM. **A** shows the measurements for Downs and G184A strains at 25 °C and after the shift up to 37 °C. **B** shows Downs shifted at 34 °C and G184A at 41 °C. The *values in parentheses* are rates of oxygen consumption expressed as µl O_2/h per mg dry weight. (Figure presented by Medoff et al. 1985 at the International Symposium on Achievements and Perspectives in Mitochondrial Research, Rosa Marina, Italy)

We have also shown that the capacity of maintaining respiration coupled at high temperature (which is specific for each given strain) is directly dependent on an optimal heat shock protein synthesis by incubating the cells at intermediate mild temperatures (see next Section; Patriarca et al. 1987; Patriarca and Maresca 1990).

Heat Shock Response

In the dimorphic human pathogenic fungus *H. capsulatum* the capacity to rapidly adapt to a new environment is directly dependent on the expression of heat shock genes. This appears to be true for parasites in general. While it has been postulated that the induction of the heat shock genes plays an active role during development (Lindquist 1986; Bond and Schlesinger 1987), their direct involvement has not yet been proven. The term "developmental activation" is used when heat shock gene activation is distinct from heat induction and occurs in cell during development (Bienz and Pelham 1987). Parasites, and dimorphic organisms like *H. capsulatum*, are particular cases of developmentally regulated heat shock gene expression: in fact, these organisms may exhibit distinct morphogenic stages and exist in multiple hosts that have different body temperatures. Temperature is therefore a crucial factor that controls phase transition but in these systems it is difficult to distinguish the effect of temperature on morphogenesis from the induction of the heat shock response.

In the mycelial phase of *H. capsulatum* the heat shock response is induced at temperatures higher than 30 °C, as shown at the protein (Shearer et al. 1987) and gene level (Caruso et al. 1987). Temperatures above 30 °C are also necessary to allow the mycelium to yeast phase transition (Maresca and Kobayashi 1989). Moreover, the pattern of the major heat shock protein synthesis differs among the tested strains and is directly dependent on the temperature used to induce transition. Heat shock protein synthesis reaches maximal activity at 34 °C in the temperature-sensitive Downs strain, while it occurs at 37 °C in the more temperature-tolerant G222B strain and in other members of this class (Shearer et al. 1987).

Our laboratory has shown that a correlation exists between the biochemical events described during the heat-activated dimorphism and the level of the heat shock response. In fact, the expression of two members of the major heat shock genes, the *hsp70* and *hsp83* genes, has been studied in two strains of *H. capsulatum* which differ in temperature sensitivity during the early period of phase transition (Caruso et al. 1987; Minchiotti et al. in prep.). Maximal transcription of these heat shock genes occurred at 34 °C in Downs, while a temperature of 37 °C was necessary to induce maximal transcription in G222B in the first 1 to 3 h after the shift-up (Fig. 4.3). Thus, the study of the heat shock gene family, its regulation, and the biological significance for the organism are important for understanding phase transition in pathogenic fungi. Heat shock genes are, in fact, among those which are induced very early during transition and may play a

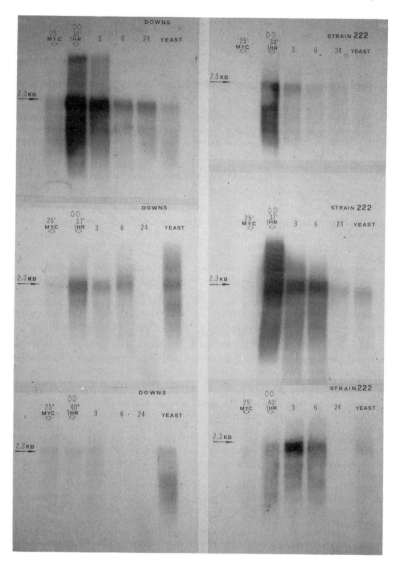

Fig. 4.3. Detection of hsp70-induced mRNA in the temperature-sensitive and less pathogenic Downs strain and in the temperature-resistant and more virulent G222B strain. Total RNA was extracted from mycelia at 25 °C, 1, 3, 6, and 24 h after temperature shock at 34, 37, and 40 °C, and from yeast cells grown for 48 h at each temperature. 24 μg of total RNA were loaded in each slot, separated on an 1% agarose gel under denaturing conditions and transferred onto Hybond −N nylon paper in 20 × SSC. The PvuII fragment containing the cloned hsp70 gene was inserted in the SmaI site of pSP65 vector and the homologous riboprobe was labeled to high specific activity (> 10^9 cpm/μg RNA). Filters were washed under stringent conditions (0.1 × SSC and 0.1% SDS) and autoradiographed for 16 h at −80 °C. (Caruso et al. 1987)

critical role in adaptation to the new environment for these organisms, rather than having a specific role in morphogenesis. At temperatures at which heat shock proteins are not induced (heat shock mRNAs are absent) the mycelial to yeast phase transition does not occur (Maresca and Kobayashi 1989).

We have found a strict correlation between temperature sensitivity (as measured by the lack of respiration and by the uncoupling of oxidative phosporylation after the temperature shift), degree of virulence, and transcription of the major heat shock genes (Patriarca et al. 1987). We have postulated that the enhanced temperature sensitivity of the Downs strain might be due to a decrease of transcription of the heat shock genes, as consequence of a low level of induction of "factors" that regulate their coordinate transcription. In fact, we showed that at 34 °C in the Downs strain the heat shock response was maximum and, as a consequence, cells did not enter the dormant phase, thus behaving like the more temperature-resistant strains. Further, under conditions in which the heat shock response was induced at high level at 37 °C, the cells had biochemical activities typical of the more thermoresistant strains and as a result they did not enter the dormant state. In fact, we have induced the heat shock proteins at high level in the Downs strain by incubating mycelial cells for 3 h at 34°C, and then measured mRNA transcription after a subsequent shift-up to 37°C. With this treatment, the Downs strain not only expressed HSP70 (Fig. 4.4) and HSP83 at normal rate at 37°C (as compared to the more virulent strains at 37°C) but cells were able to continue to respire and oxygen transport remained coupled to oxidative phosphorylation (Patriarca et al. 1987; Patriarca and Maresca 1990). Similar results were obtained when more temperature-resistant strains were used and in this case cells were shifted to 37 °C before a further shift to temperature ranging between 39 and 41 °C (Patriarca and Maresca 1990; Minchiotti et al. in prep.) From these data it is clear that the induction of sufficient heat shock protein is required for the maintenance of oxidative phosphorylation which by allowing ATP synthesis maintains normal RNA and protein synthesis and this in turn allows the cells to proceed towards phase transition.

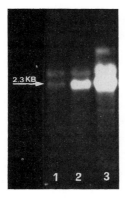

Fig. 4.4. Detection of HSP70-induced mRNA in Downs cells. Total RNA was extracted from mycelia at 25°C (*1*), from mycelia incubated for 5 h at 34°C (*2*) and then shifted for 1 h at 37°C (*3*). The same riboprobe described in the legend of Fig. 3 was used in the hybridization reaction at the same conditions. (Figure presented by Patriarca et al. 1987 at the 100th Anniversary Meeting of the Institute Pasteur, Paris.)

Another important feature in the adaptation of *H.capsulatum* to the new environmental temperature is the maturation of the hsp83 mRNA. In *Drosophila* cells the hsp83 gene, which contains one intervening sequences, is properly spliced in conditions of mild heat shock, while its maturation (along with other intron-containing genes) is blocked under severe heat shock (Yost and Lindquist 1986). These authors have demonstrated that the heat-induced block in splicing is general, by showing that the maturation of Adh (alcohol dehydrogenase) transcripts is equally sensitive to heat.

The *H. capsulatum* hsp83 gene contains two introns which are properly spliced also at extreme temperatures during the mycelial to yeast transition at 37 °C for Downs and 41 °C for G222B (Minchiotti et al. in prep.). Therefore this situation does not resemble that found in *Drosophila* and humans (Bond 1988) under experimental conditions. This is not surprising because, in nature, eukaryotic systems generally undergo a gradual increase in temperature, only rarely experiencing sudden large increases in temperature which must be lowered to physiological level to ensure survival. On the contrary, heat shock is not an artificial situation in dimorphic organisms after infection and the splicing process must be maintained active in order to allow the organism to undergo a morphogenic transition to the parasitic phase.

Conclusion

The morphological differentiation process in *Histoplasma capsulatum* is intimately correlated to a new environment and to a heat shock phenomenon. An example of this capacity to adapt to the new environmental conditions is activation of heat shock genes during induction of the yeast phase, although it is reasonable to postulate that such adaptation and morphologic response require a more complex specific mechanism at the gene level. Groups of genes should be coordinately activated for adaptation to the new conditions together with specific sets of regulatory and phase transition genes. How temperature intervenes in gene expression is not yet clear. The analysis of early genes coding for regulatory functions, as well as studies of mutants and transfection analysis of *in vitro*-generated mutations of putative regulatory genes, seem the best route now to an understanding of the regulatory induction of genetic pathways that leads to phase transition, and possibly to the mechanisms that induce virulence in dimorphic pathogenic fungi.

Acknowledgements

This work was supported by a Contract from the Commission of the European Communities, TSD2-0132-I and a Contract from Ministero della Sanità, Progetto AIDS, #4204 22.

References

Alberts B, Bray D, Lewis J, Raff M, Roberts K, Watson D (1983) Molecular Biology of the Cell. Garland, New York

Ashburner M, Bonner JJ (1979) The induction of gene activity in *Drosophila* by heat shock. Cell 17:241–254

Bienz M, Pelham RHB (1987) Mechanism of heat shock gene activation in higher eukaryotes. Adv Genet 24:31–72

Bond U (1988) Heat shock but not other stress inducers leads to the disruption of a sub-set of snRNPs and inhibition of *in vitro* splicing in HeLa cells. Embo J 7:3509–3518

Bond U, Schlesinger MJ (1987) Heat shock proteins and development. Adv Genet 24:1–29

Caruso M, Sacco M, Medoff G, Maresca B (1987) Heat shock 70 gene is differentially expressed in *Histoplasma capsulatum* strains with different levels of thermotolerance and pathogenicity. Mol Microbiol 1:151–158

Craig EA (1985) The heat shock response. Crit Rev Biochem 18:239–280

Lambowitz AM, Kobayashi GS, Painter A, Medoff G (1983) Possible relationship of morphogenesis in the pathogenic fungus *Histoplasma capsulatum* to heat shock response. Nature (Lond) 303:806–808

Lindquist S (1986) The heat shock response. Annu Rev Biochem 55:1151–1191

Maresca B, Kobayashi GS (1989) Dimorphism in *Histoplasma capsulatum*: a model for the study of cell differentiation in pathogenic fungi. Microbiol Rev 53:186–209

Maresca B, Lambowitz AM, Kobayashi GS, Medoff G (1979) Respiration in the yeast and mycelial phases of *Histoplasma capsulatum*. J Bacteriol 138:647–649

Maresca B, Lambowitz AM, Kumar BV, Grant GA, Kobayashi GS, Medoff G (1981) Role of cysteine oxidase in regulating morphogenesis and mitochondrial activity in the dimorphic fungus *H.capsulatum*. Proc Natl Acad Sci USA 78:4596–4600

McVeigh I, Houston WE (1972) Factors affecting mycelial to yeast phase conversion and growth of the yeast phase of *Histoplasma capsulatum*. Mycopathol Mycol Appl 47:135–151

Medoff G, Lambowitz AM, Painter A et al. (1985) Sulfhydryl induced respiratory shunt pathways and morphogenesis in several dimorphic pathogenic fungi. Int Sym Achievements and Perspectives in Mitochondrial Research. Rosa Marina, Italy, Sep 2–6

Medoff G, Maresca B, Lambowitz AM et al. (1986) Correlation between pathogenicity and temperature sensitivity in different strains of *Histoplasma capsulatum*. J Clin Invest 78:1638–1647

Mues GI, Munn TZ, Raeses JD (1986) A human gene family with sequence homology to *Drosophila melanogaster hsp70* heat shock genes. J Biol Chem 261:874–877

Patriarca EJ, Maresca B (1990) Mitochondrial activity and heat shock response during morphogenesis in pathogenic fungus *Histoplasma capsulatum*. Mol Microbiol (submitted)

Patriarca EJ, Sacco M, Carratù L, Minchiotti G, Maresca B (1987) Gene expression during phase transition in the dimorphic pathogenic fungus *Histoplasma capsulatum*. Centenary symp Mol Biol Infect Des Inst Pasteur. Oct 5–9 1987, p 145

Scherr GH (1957) Studies on the dimorphism of *Histoplasma capsulatum*. The role of -SH groups and incubation temperature. Exp Cell Res 12:92–107

Schwarz L (1981). Histoplasmosis. Praeger Science Press, New York

Shearer G, Birge C, Yuckenberg PD, Kobayashi GS, Medoff G (1987) Heat shock proteins induced during the mycelial-to-yeast transitions of strains of *Histoplasma capsulatum*. J Gen Microbiol 133:3375–3382

Yost HJ, Lindquist S (1986) RNA splicing is interrupted by heat shock and is rescued by heat shock protein synthesis. Cell 45:185–193

CHAPTER 5

The Cellular Functions of Chaperonins

Anthony A. Gatenby, Gail K. Donaldson, Pierre Goloubinoff,
Robert A. LaRossa, George H. Lorimer, Thomas H. Lubben,
Tina K. Van Dyk and Paul V. Viitanen

Molecular Biology Division, Central Research and Development Department, Experimental
Station, E.I. DuPont de Nemours & Co., Wilmington, DE 19880-0402, USA

Introduction

It is apparent that a major sub-set of heat shock proteins assist other polypeptides to maintain, or assume, a conformation required for their correct assembly into biologically active structures (Georgopoulos et al. 1973; Kochan and Murialdo 1983; Goloubinoff et al. 1989; Cheng et al. 1989; Ostermann et al. 1989; Bresnick et al. 1989) or localization (Deshaies et al. 1988; Chirico et al. 1988; Zimmermann et al. 1988; Bochkareva et al. 1988; Lecker et al. 1989). This group of proteins function as molecular chaperones, and they have been defined as proteins which assist the assembly of some oligomeric proteins, but are not components of the final structure (Ellis 1987; Ellis et al. 1989; Ellis and Hemmingsen 1989). One distinct group of related molecular chaperones are found in prokaryotes, mitochondria, and plastids, and are called chaperonins (Hemmingsen et al. 1988). In this chapter we outline the discovery and characterization of chaperonins in prokaryotic and eukaryotic organisms, and also describe recent data that show that these proteins have an important role in protein folding in cells.

Identification of Chaperonins in Organelles and Their Role in Eukaryotic Cells

Chloroplastic Chaperonins

The chloroplast enzyme D-ribulose-1,5-bisphosphate carboxylase/oxygenase (Rubisco) is a hexadecameric structure in which eight nuclear encoded small subunits are appended, in two clusters of four, to the two poles of a core of eight chloroplast encoded large subunits (for reviews see Gutteridge and Gatenby 1987; Andrews and Lorimer 1987). The first indication that the biosynthesis of Rubisco is not a simple self-assembly process and might require additional protein components was demonstrated by Barraclough and Ellis (1980). When isolated pea (*Pisum sativum*) chloroplasts are incubated in a suitable medium

Stress Proteins
Schlesinger, Santoro, Garaci (Eds.)
© Springer-Verlag Berlin Heidelberg 1990

with radioactive amino acids, the synthesis of chloroplast proteins can be observed, the most abundant of which is the large subunit of Rubisco (Blair and Ellis 1973). Electrophoresis on nondenaturing polyacrylamide gels reveals that whereas some of these newly synthesized radioactive large subunits are indeed assembled into Rubisco holoenzyme, a large proportion are not assembled and become associated with a protein of subunit molecular weight of 60 000 (polypeptide 60) (Barraclough and Ellis 1980). These authors noted that the protein containing the nonassembled large subunits, together with the unlabeled 60 000 subunits, has a molecular weight of greater than 600 000. The results of time course experiments following radiolabeling show that as radioactive large subunits become assembled into the Rubisco holoenzyme, the radioactivity in the 600 000 protein declines. These observations raised the possibility that nascent large subunits are specifically associated with polypeptide 60 prior to assembly, and that this complex is an obligatory intermediate in the assembly of Rubisco. Polypeptide 60 was subsequently referred to as the large subunit-binding protein, and it was suggested that it is likely to be a nuclear gene product because although it can be labeled in vivo with radioactive amino acids, it is not labeled in isolated chloroplasts actively synthesizing organelle encoded proteins (Ellis 1981).

Experiments by Roy et al. (1982) also demonstrate that large subunits synthesized in vivo or in organello can be recovered from intact chloroplasts in the form of two different sedimentation complexes of 7S and 29S on sucrose gradients. The 29S complex contains unassembled Rubisco large subunits and large subunit binding protein, and it is identical to the complex of molecular weight 600 000 observed by Barraclough and Ellis (1980). The 7S complex is probably a dimer of large subunits. When chloroplasts are incubated in the light it is found that the newly synthesized large subunits present in both the 7S and 29S complexes disappear and are subsequently found in the assembled 18S Rubisco molecule (Roy et al. 1982). Bloom et al. (1983) showed that light can promote the posttranslational assembly of large subunits into Rubisco, and this assembly is accelerated in chloroplast extracts by the addition of ATP, but the 29S large subunit binding protein oligomer is maintained. In the presence of Mg^{2+}, however, ATP causes dissociation of the 29S complex whereas a nonhydrolyzable analog of ATP has no effect. A complex set of reactions was proposed by Bloom et al. (1983) that requires nucleotides, Mg^{2+}, and putative intermediates in the assembly of Rubisco holoenzyme.

Data published by Gatenby et al. (1988) and Ellis and van der Vies (1988) revealed that not only does the large subunit of Rubisco interact with the large subunit binding protein, but imported radiolabeled small subunits of Rubisco are also associated with it (Fig. 5.1b). In addition, foreign prokaryotic large subunits of Rubisco imported into pea chloroplasts will also associate with the large subunit binding protein (Fig. 5.1a) (Gatenby et al. 1988). The name of the protein was thus changed to Rubisco subunit binding protein to reflect these new observations (Ellis and van der Vies 1988). The Rubisco subunit binding protein has been purified from pea (Hemmingsen and Ellis 1986) and barley (*Hordeum vulgare*) (Musgrove et al. 1987). It has a molecular weight of 720 000

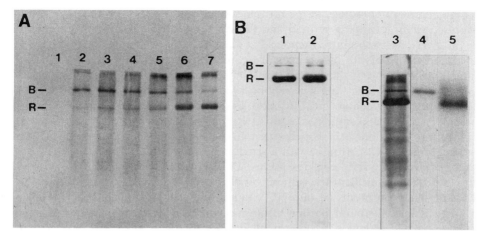

Fig. 5.1. Association of imported large (**A**) or small (**B**) subunits of Rubisco with the chloroplast chaperonin. **A** Import time course of Rubisco large subunits fused to a transit peptide, labeled with ^{35}S-methionine, and incubated with isolated pea chloroplasts. Aliquots of chloroplasts were removed from the import reaction at various times, treated with protease, and a stromal fraction prepared. An identical amount of protein from each time point was electrophoresed on a 6% nondenaturing polyacrylamide gel, followed by fluorography. Time points were 0 (lane 1); 5 (lane 2); 10 (lane 3); 20 (lane 4); 30 (lane 5); 60 (lane 6) and 90 min (lane 7). The position of the chloroplast chaperonin is marked *B*, and that of assembled Rubisco is marked *R*. **B** In lanes 1 and 2 ^{35}S-methionine-labeled Rubisco small subunits were imported into isolated pea chloroplasts. Samples were prepared and electrophoresed as in panel A, stained with Coomassie blue, and fluorographed. Lane *1* shows the stained protein and land *2* is the fluorograph of the same sample. Notice that although most imported small subunits have assembled into Rubisco holoenzyme (*R*), some are found associated with the chloroplast chaperonin oligomer (*B*). In lanes *3–5* unlabeled stroma was electrophoresed on a 5% nondenaturing polyacrylamide gel and either stained with Coomassie blue (lane *3*), or immuno-blotted using antiserum against chloroplast chaperonin (lane *4*), or Rubisco (lane *5*), followed by ^{125}I-protein A detection. This allowed identification of the relative positions of the chaperonin (*B*) and Rubisco (*R*) oligomers. (Gatenby et al. 1988)

and is composed of two types of subunits of 60000 and 61000 (Hemmingsen and Ellis 1986). The larger polypeptide is referred to as the α subunit, the smaller is the β subunit, and the two subunits are immunologically distinct, show different partial protease digestion patterns, and have different amino-terminal sequences (Musgrove et al. 1987). Antisera raised against the Rubisco subunit binding protein were used to show that a high molecular weight form of the polypeptide is immunoprecipitated from products of pea polysomes translated in a wheat germ cell-free translation system, thus confirming that the protein is synthesized on cytoplasmic ribosomes.

The availability of highly specific antisera raised against purified Rubisco subunit binding protein from pea chloroplasts facilitated the isolation of cDNA clones for the α subunit (Hemmingsen et al. 1988). The clones from castor bean (*Ricinus communis*) and wheat (*Triticum aestivum*) α subunits encode proteins with predicted amino-terminal sequences that closely match the determined amino-terminus of the purified mature α polypeptide of the Rubisco subunit binding protein from both wheat and pea. The wheat and castor bean sequences are 80% identical at the amino acid level. A remarkable finding, however, is a high degree of homology at the protein sequence level between the chloroplast Rubisco subunit binding protein and the *E. coli* GroEL protein (Hemmingsen et al. 1988). With the introduction of some gaps, 46% of the residues are identical and many of the differences are conservative substitutions. As discussed later, one function of the GroEL protein is to assist in bacteriophage morphogenesis, during which time it is required for the earliest assembly step that has been detected (Kochan and Murialdo 1983). Therefore, the posttranslational assembly of some complex structures, Rubisco, and certain bacteriophages in chloroplasts and bacteria appears to make use of related proteins during the assembly process. A protein with a structure similar to the *E. coli* GroEL protein has been purified from pea leaves (Pushkin et al. 1982), and in retrospect this is probably the Rubisco subunit binding protein. It was proposed that this class of related proteins be termed "chaperonins" (Hemmingsen et al. 1988).

The Rubisco subunit binding protein has now undergone a further change of name, and we now propose to call it chloroplast chaperonin 60 (cpn60) in view of its size on SDS polyacrylamide gels (60 kDa), and to recognize the class of proteins (chaperonins) to which it belongs. An additional reason for discarding the name Rubisco subunit binding protein is the observation that the protein does have a more general role in the chloroplast other than simply Rubisco assembly (Lubben et al. 1989). The previously observed association of only the large subunits of Rubisco with cpn60 in isolated chloroplasts may just be a reflection of the prodigious amounts of large subunits synthesized in the isolated organelles, compared to other proteins being synthesized. An alternative way to examine the interaction of a range of proteins with chloroplast cpn60 is to synthesize radiolabeled proteins in an in vitro cell-free translation system. If the proteins are designed to have an amino-terminal chloroplast transit peptide, they can be imported into the isolated organelles and their association with cpn60 measured. With this approach, we have examined the interaction of cpn60 with nine proteins imported into isolated pea chloroplasts. The proteins used for these experiments were the bacterial proteins pre-β-lactamase (monomeric) and chloramphenicol acetyltransferase (trimeric), and the plant proteins *Silene pratensis* ferredoxin (monomeric), pea superoxide dismutase (dimeric), and alfalfa (*Medicago sativa*) glutamine synthetase (octomeric). Two membrane proteins, the β-subunit of the thylakoid ATP synthase from maize (*Zea mays*) chloroplasts and the light-harvesting chlorophyll a/b protein (LHCP) from *Petunia hybrida* were also imported. The *Anacystis nidulans* large and pea small subunits of Rubisco were imported as controls, since their interaction with cpn60 is known (Gatenby et al. 1988; Ellis and van der Vies 1988). The transit peptide of the small subunit of Rubisco from soybean (*Glycine max*) was used in many of these fusion proteins to direct import, but some proteins used their natural transit peptide (ferredoxin, superoxide dismutase, LHCP). After the radiolabeled proteins were imported, the chloroplasts were lysed and a stromal fraction obtained. Following various treatments the samples were analyzed by electrophoresis on denaturing or nondenaturing polyacrylamide gels.

When untreated extracts were examined, all radiolabeled imported proteins, except ferredoxin and superoxide dismutase, were found to comigrate with the oligomeric form of cpn60 on non-denaturing gels (Lubben et al. 1989). To determine whether proteins associated with cpn60 rather than merely comigrating with it, advantage was taken of the observation that cpn60 oligomers dissociate in the presence of ATP, in a reaction that requires ATP hydrolysis (Bloom et al. 1983; Hemmingsen and Ellis 1986; Roy et al. 1988). When $MgCl_2$ and ATP were added to stromal extracts made from chloroplasts which contained imported proteins, cpn60 oligomers dissociated and the imported radiolabeled polypeptides were no longer found in that part of the gel. Following this treatment, some of the radiolabeled protein migrated near the top of nondenaturing gels as a high molecular weight complex. No dissociation was observed when the nonhydrolyzable analog β,γ-methyleneadenosine 5′-triphosphate was used instead of ATP. From these studies we conclude that those proteins which comigrated with the oligomeric cpn60 were associated with it.

The interaction of several different imported proteins with cpn60 suggests a general role for this oligomer at some point in the import, folding, or assembly pathway of chloroplast proteins, particularly if proteins are wholly or partially unfolded during import and subsequently require refolding. If binding of these imported polypeptides to cpn60 represents an early event in folding, then the interaction of such a diverse group of proteins with the chloroplast chaperonin implies that a common and very general recognition motif may be present in these proteins. Chaperonins are not considered to be part of the final structure (Hemmingsen et al. 1988), and apparently do not bind to completely folded proteins. Therefore, this motif may only be accessible in polypeptides entering the chloroplast in an incompletely folded state. The ability to recognize common sequential or transiently exposed structural features of the partially folded imported protein could account for the broad specificity of interactions. This type of protein interaction may also explain the association of cpn60 with Rubisco large subunits synthesized within isolated chloroplasts (Barraclough and Ellis 1980; Roy et al. 1982). Elongating polypeptide chains emerging from ribosomes may present features during translation, or after release in an unfolded or partially folded form, that are available for interaction with cpn60 in a manner comparable to the interaction with polypeptides entering the organelle. Because both plastids and mitochondria contain their own chaperonins, these proteins may therefore have a crucial role in mediating the posttranslational folding and assembly of many imported or organelle-encoded proteins.

Mitochondrial Chaperonins

The presence of chaperonins in mitochondria was described by McMullin and Hallberg (1988). A *Tetrahymena thermophila* mitochondrial protein of molecular weight 58 000 is selectively synthesized during heat shock, and displays antigenic similarity to a protein of approximately the same size found in mitochondria from a wide range of species, including *Saccharomyces cerevisiae*, *Xenopus laevis*, maize, and human cells. These related proteins form oligomeric complexes with comparable sedimentation characteristics and identical morphologies (McMullin and Hallberg 1988; Prasad and Hallberg 1989). An analogous protein is also present in *Neurospora crassa* mitochondria (Hutchinson et al. 1989). In addition, a protein is present in *E. coli* that cross-reacts with the *T. thermophila* mitochondrial antigen. Based on its characteristic morphology, the antigenically related protein in *E. coli* has been identified as the GroEL protein. The resemblance between the eukaryotic mitochondrial proteins and *E. coli* GroEL was confirmed by prediction of protein sequences from isolated genes (Reading et al. 1989; Jindal et al. 1989), and directly by amino acid sequence homology (Waldinger et al. 1988). These authors reported that not only are homologies found between the mitochondrial heat shock protein (HSP60) and GroEL, but they are also observed between the chloroplast cpn60 and mitochondrial HSP60.

An important role for HSP60 during the biogenesis of *S. cerevisiae* mitochondria was demonstrated by Cheng et al. (1989), who found that this heat shock protein is required for the assembly into oligomeric complexes of proteins imported into the mitochondrial matrix. Temperature-sensitive lethal *S. cerevisiae* mutants were examined for a phenotype resulting from a deficiency in the assembly of mitochondrial proteins. A nuclear mutation, designated *mif4*, was isolated that prevents the assembly of oligomeric protein complexes of F_1-ATPase, cytochrome b_2 and the Rieske Fe/S protein of complex III. A wild-type copy of the *Mif4* gene was isolated by transforming the mutant strain with a library containing yeast genomic DNA fragments and selecting for growth at 37 °C. One of the rescuing plasmids encoded the HSP60 protein. By integration of a wild-type copy of hsp60 into the hsp60 locus of mutant cells, followed by mating with a wild-type strain, sporulation, and tetrad analysis, it was confirmed that the *mif4* mutation is in the hsp60 structural gene. It was therefore concluded that polypeptides entering the mitochondrial matrix space from the cytosol require the *Mif4* gene production HSP60, a matrix protein, for assembly and/or for further sorting events (Cheng et al. 1989). As discussed in the previous section, both chloroplasts and mitochondria probably contain chaperonins to assist correct folding and assembly of proteins that are imported into the organelles. This view is supported by data from Ostermann et al. (1989), who show that folding of imported proteins occurs at the surface of HSP60 in an ATP-mediated reaction. They further propose that HSP60 catalyses protein folding. Additionally, chaperonins in mitochondria may assist in protein folding and assembly of proteins synthesized within the organelle.

Prokaryotic Chaperonins

Protein Folding and Assembly

The earliest evidence concerning the involvement of chaperonins with posttranslational events was the demonstration that the *E. coli* cpn60 (GroEL) and cpn10 (GroES) proteins are essential for phage head morphogenesis (Sternberg

1973; Georgopoulos et al. 1973; Zweig and Cummings 1973). The GroE proteins are required for an early step in morphogenesis, which is assembly of bacteriophage λ preconnectors (Kochan and Murialdo 1983; Kochan et al. 1984). These are ring-shaped oligomeric structures composed of 12 subunits of a phage-encoded protein gpB, upon which head shell assembly occurs. Both the GroEL and GroES proteins are involved in DNA replication in *E. coli* (Fayet et al. 1986; Jenkins et al. 1986). Recent studies have shown that GroEL binds unfolded cytoplasmic and secretory proteins, and that it may maintain precursor proteins in a conformation which is competent for membrane translocation (Bochkareva et al. 1988; Lecker et al. 1989).

The close homology between *E. coli* GroEL and chloroplast cpn60 proteins (Hemmingsen et al. 1988) led us to consider experiments in which the influence of *groE* products on Rubisco assembly in vivo could be examined (Goloubinoff et al. 1989). The expression of the Rubisco large and small subunit genes of the cyanobacterium *Anancystis nidulans* in *E. coli* results in synthesis and assembly of an active Rubisco holoenzyme (Gatenby et al. 1985; van der Vies et al. 1986). To demonstrate that GroE proteins influence the assembly of *A. nidulans* Rubisco in *E.coli*, the level of cellular GroE proteins was increased by cloning the *groE* genes on a multicopy plasmid. The *groE* genes were cloned on a chloramphenicol-resistant plasmid (pGroESL), made from a vector (pTG10) that is compatible with an ampicillin-resistant plasmid that directs the synthesis of Rubisco subunits (pANK1). The presence of the plasmid encoding the *groE* operon (pGroESL) in *E. coli* increases the amount of GroEL protein from its normal level of 1–2% of cell protein, to about 10% (Fig. 5.2b, d).

Cells containing the plasmid encoding the *A. nidulans rbcL* and *rbcS* genes (pANK1), and synthesizing wild-type levels of GroE proteins, produce active Rubisco enzyme that represents 0.7% of cell protein during logarithmic growth (Fig. 5.2, sample 2). An increase in the cellular concentration of GroE proteins, directed by plasmid pGroESL, leads to a substantial increase in Rubisco activity (Fig. 5.2a, sample 3). An increase in the amount of assembled Rubisco holoenzyme is also apparent when cell extracts are examined by electrophoresis on nondenaturing gels (Fig. 5.2b, sample 3). The assembled Rubisco in GroE overproducing strains represents 5–6% of cell protein. Despite the almost ten fold variation in active assembled holoenzyme between wild-type and GroE overproducing strains, each strain contained comparable amounts of large subunit polypeptide. Therefore, the increased amount of assembled and active Rubisco arising from overproducing the GroE proteins is not due to increased expression of the Rubisco genes, or increased protein stability. Instead, the GroE proteins influence the posttranslational assembly of Rubisco polypeptides into a functional holoenzyme.

It was also found that *groES* and *groEL* defective strains which prevent bacteriophage morphogenesis, will inhibit the assembly of hexadecameric Rubisco. Introduction of the pGroESL plasmid into these defective strains restored the ability to assemble pANK1 encoded large and small subunits of Rubisco. By using appropriate combinations of mutant strains and plasmids it is apparent that the cellular concentration of GroE proteins influence not only

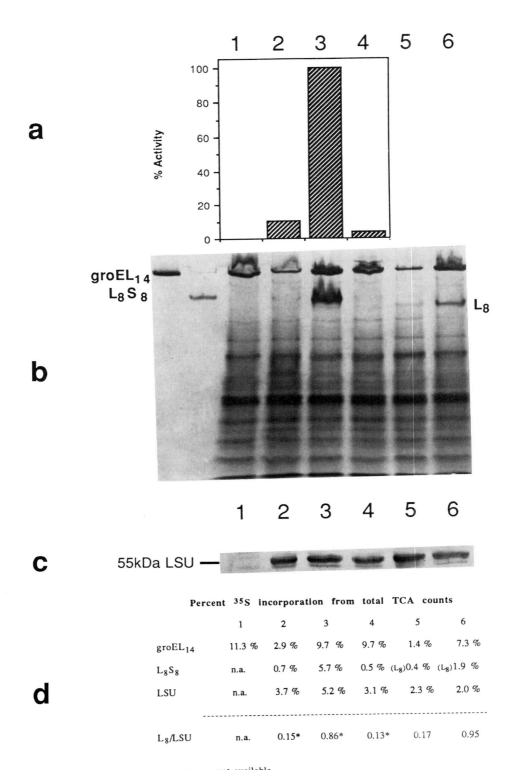

a

% Activity

1 2 3 4 5 6

100

80

60

40

20

0

b

groEL$_{14}$
L$_8$S$_8$

L$_8$

c

1 2 3 4 5 6

55kDa LSU —

d

Percent ^{35}S incorporation from total TCA counts

	1	2	3	4	5	6
groEL$_{14}$	11.3 %	2.9 %	9.7 %	9.7 %	1.4 %	7.3 %
L$_8$S$_8$	n.a.	0.7 %	5.7 %	0.5 %	(L$_8$)0.4 %	(L$_8$)1.9 %
LSU	n.a.	3.7 %	5.2 %	3.1 %	2.3 %	2.0 %
L$_8$/LSU	n.a.	0.15*	0.86*	0.13*	0.17	0.95

n.a. - not available
* - Predicted L$_8$ values from L$_8$S$_8$ values

the assembly of hexadecameric Rubisco, but also the assembly of both the *A. nidulans* large subunit core octomer encoded by pDB53 (Fig. 5.2b, samples 5 and 6) and the structurally simpler *Rhodospirillum rubrum* Rubisco dimer (Goloubinoff et al. 1989). The significance of this observation is that recent structural studies of the L_8S_8 holoenzyme have revealed that the L_8 core is composed of four *R. rubrum*-like dimers arranged about a fourfold axis, the L_2 dimer being the basic structural and catalytically functional motif (Schneider et al. 1986; Champman et al. 1988; Andersson et al. 1989). This suggests that assembly of L dimers may be an essential intermediary stage in the formation of the L_8 core, and that therefore the influence of GroE proteins on the basic common step of formation of dimers can control the subsequent assembly of the more complex forms of Rubisco. Recent experiments have investigated the role of GroE proteins in dimer formation in vitro, and these have demonstrated that the *groE* products are directly involved in protein folding (PG, JT Christeller, AAG, GHL, submitted).

Suppression of Heat-Sensitive Mutations

Experiments reviewed thus far illustrate that in eukaryotic organelles the chaperonins can interact with a diverse range of imported proteins, and in prokaryotes they are involved in bacteriophage morphogenesis and the assembly of the Rubisco holoenzyme. The assembly of bacteriophages and Rubisco cannot be the sole role of chaperonins in bacteria, because these two examples represent respectively a pathogen and a foreign expressed protein. In an attempt to examine the extent to which bacterial proteins interact with chaperonins, we have tested the ability of the *E. coli groE* operon to suppress mutations in both *E. coli* and *Salmonella typhimurium* (Van Dyk et al. 1989). We proposed that overexpression of the *groE* products might correct growth defects by forcing either the folding or assembly of recalcitrant, mutant polypeptides.

Initially, heat-sensitive mutations in the major *ilv* operon in *S. typhimurium* encoding enzymes of branched chain amino acid biosynthesis were examined. The auxotrophic requirements of one mutation in *ilvGM* and two mutations in *ilvE*, encoding the multimeric enzymes acetolactate synthase II and transaminase B, respectively, are suppressed by a multicopy plasmid encoding the *groE* operon. To extend these observations, the response of a large number of mutations to multiple copies of *groE* was examined. Many, but not all, heat-sensitive *hisD*, *hisC* and *hisB* alleles are suppressed by *groE* overexpression.

◁──

Fig. 5.2. In vivo assembly of Rubisco in *E. coli*. A *recA56* isolate of strain MC1061 was cotransformed with the following plasmids: *1* pUC18 and pGroESL; *2* pANK1 and pTG10; *3* pANK1 and pGroESL; *4* pANK1 and pGroEL ; *5* pDB53 and pTG10; *6* pDB53 and pGroESL. Plasmids are described in more detail in the text. **a** Rubisco activity in the cell lysates. **b** Nondenaturing 7% polyacrylamide gel electrophoresis of lysate proteins, stained with Coomassie blue. Purified GroEL and spinach L_8S_8 Rubisco are shown on the *two left lanes*. **c** SDS-polyacrylamide gel electrophoresis of cell lysates, revealed by immunodetection with an antiserum raised against *A. nidulans* Rubisco large subunits. **d** ^{35}S-counts incorporated in protein complexes from native and SDS-polyacrylamide gels as in **b** and **c**, respectively as a percentage of total TCA counts. *n.a.* not available; *asterisk* predicted L_8 values from L_8S_8 values; *LSU* large subunit. (Goloubinoff et al. 1989)

These genes encode histidinol dehydrogenase, histidinol-phosphate aminotrans-
ferase and the bifunctional protein imidazoleglycerol phosphate dehydratase:
histidinol phosphate phosphatase, respectively. Suppression of cold-sensitive
and temperature-independent alleles has not been observed in the *his* system.
Many extensively characterized heat-sensitive protein-folding mutations of
Salmonella phage P22 gene *9*, encoding the trimeric tailspike, prevent plaque
formation on wild-type *S. typhimurium* at the restrictive temperature (Yu and
King 1984). Several of these mutations are suppressed when the phage is plated
on a host containing the *groE* multicopy plasmid, allowing plaque formation to
occur. The presence of the *groE* plasmid also suppresses heat-sensitive growth
and secretion defects caused by the *secA51* and *secY24* alleles of *E. coli*. Thus,
mutations causing alterations in enzymes, structural proteins, and the protein
translocation apparatus can all be suppressed.

Overexpression of the *groE* operon in *E. coli* is known to suppress the heat-
sensitive phenotype of several *dnaA* alleles (Fayet et al. 1986; Jenkins et al.
1986), and it had been inferred that suppression of heat-sensitive mutations is
confined to *dnaA* alleles reflecting an interaction between the *groE* products
and a dnaA protein aggregate at the replication origin. Suppression is also observed of an *E. coli ssb-1* mutation by an allele of *groEL* (Ruben et al. 1988).
Our observations that multiple copies of the *groE* operon suppress mutations in
genes encoding several diverse proteins suggests a general role for the *groE*
operon products (GroES and GroEL). Both *groE* genes are needed to mediate the suppression in the mutations we have examined, a requirement that is
also necessary for suppression of *dnaA* alleles (Fayet et al. 1986; Jenkins et al.
1986), bacteriophage morphogenesis (Tilly et al. 1981), and Rubisco assembly
(Goloubinoff et al. 1989). We interpret this *groE*-mediated genetic suppression
to imply a direct interaction of the *groE* products with the mutant proteins. If
mutant proteins are more rapidly directed down unproductive pathways of
either irreversible degradation or aggregation, then increasing the cellular concentration of the GroE proteins may trap the unstable mutated proteins and
enable them to fold correctly. Such interactions may reflect a normal association
of the GroE proteins with the translation products of many genes, including
wild type versions of the suppressed alleles. Other alternative explanations, for
example alterations of translational fidelity, expression levels, induced translational misreading, codon specific mechanisms of informational suppression,
or an anti-proteolytic role for the GroE proteins appear to be inconsistent with
available data (Van Dyk et al. 1989).

Concluding Remarks

From the data reviewed and presented here it is clear that not only are chaperonins ubiquitous, but they are also involved in a diverse array of cellular functions. The bacterial and mitochondrial chaperonins are induced as stress response
proteins, but in addition they are constitutively expressed during normal cell

growth and are essential for cell viability (Fayet et al. 1989; Cheng et al. 1989). Although it is commonly believed that the amino acid sequence of a protein is sufficient to specify its correct folding into a final active conformation (Anfinsen 1973), it is now known that in some instances this process is stimulated by chaperonins (Cheng et al. 1989; Goloubinoff et al. 1989; Ostermann et al. 1989). Recent published (Ostermann et al. 1989) and unpublished (PG, JT Christeller, AAG, GHL, submitted) experiments demonstrate that the chaperonins are involved directly in protein folding.

If the posttranslationally assisted folding of proteins by chaperonins is common, as suggested by our chloroplast import experiments and suppression of heat-sensitive mutations, then these auxiliary proteins have a critical role in cells. Their folding activity may be important, not only for proteins imported into organelles or released from ribosomes with an incomplete three-dimensional structure, but they could also function in anti-folding to maintain polypeptides in a translocation competent conformation until they have entered the secretory pathway in bacteria. A major challenge now is to understand the molecular mechanism responsible for the interaction and function of chaperonins with such a wide range of proteins and cellular processes.

References

Andersson I, Knight S, Schneider G et al. (1989) Crystal structure of the active site of ribulose-bisphosphate carboxylase. Nature (Lond) 337:229–234

Andrews TJ, Lorimer GH (1987) Rubisco: structure, mechanisms, and prospects for improvement. In: Hatch MD, Boardman NK (eds) The biochemistry of plants, Vol 10. Academic Press, Lond New York, p 131

Anfinsen CB (1973) Principles that govern the folding of protein chains. Science 181:223–230

Barraclough R, Ellis RJ (1980) Protein synthesis in chloroplasts. IX. Assembly of newly-synthesized large subunits into ribulose bisphosphate carboxylase in isolated pea chloroplasts. Biochim Biophys Acta 608:19–31

Blair GE, Ellis RJ (1973) Protein synthesis in chloroplasts. I. Light-driven synthesis of the large subunit of Fraction I protein by isolated pea chloroplasts. Biochim Biophys Acta 319:223–234

Bloom MV, Milos P, Roy H (1983) Light-dependent assembly of ribulose 1,5-bisphosphate carboxylase. Proc Natl Acad Sci USA 80:1013–1017

Bochkareva ES, Lissin NM, Girshovich AS (1988) Transient association of newly synthesized unfolded proteins with the heat shock GroEL protein. Nature (Lond) 336:254–257

Bresnick EH, Dalman FC, Sanchez ER, Pratt WB (1989) Evidence that the 90-kDa heat shock protein is necessary for the steroid binding conformation of the L cell glucocorticoid receptor. J Biol Chem 264:4992–4997

Chapman MS, Suh SW, Curmi PMG, Cascio D, Smith WW, Eisenberg DS (1988) Tertiary structure of plant rubisco: domains and their contacts. Science 241:71–74

Cheng MY, Hartl FU, Martin J et al. (1989) Mitochondrial heatshock protein HSP60 is essential for assembly of proteins imported into yeast mitochondria. Nature (Lond) 337:620–625

Chirico WJ, Waters MG, Blobel G (1988) 70K heat shock proteins stimulate protein translocation into microsomes. Nature (Lond) 332:805–810

Deshaies RJ, Koch BD, Werner-Washburne M, Craig EA, Schekman R (1988) A subfamily of stress proteins facilitates translocation of secretory and mitochondrial precursor polypeptides. Nature (Lond) 332:800–805

Ellis RJ (1981) Chloroplast proteins: synthesis, transport and assembly. Ann Rev Plant Physiol 32:111–137

Ellis RJ (1987) Proteins as molecular chaperones. Nature (Lond) 328:378–379

Ellis RJ, Hemmingsen SM (1989) Molecular chaperones: proteins essential for the biogenesis of some macromolecular structures. Trends Biochem Sci 14:339–342

Ellis RJ, van der Vies SM (1988) The rubisco subunit binding protein. Photosynth Res 16:101–115

Ellis RJ, van der Vies SM, Hemmingsen SM (1989) The molecular chaperone concept. Biochem Soc Symp 55:145–153

Fayet O, Louarn J-M, Georgopoulos C (1986) Suppression of the *Escherichia coli* dnaA46 mutation by amplification of the groES and groEL genes. Mol Gen Genet 202:435–445

Fayet O, Ziegelhoffer T, Georgopoulos C (1989) The GroES and GroEL heat shock gene products of *Escherichia coli* are essential for bacterial growth at all temperatures. J Bact 171:1379–1385

Gatenby AA, van der Vies SM, Bradley D (1985) Assembly in *E. coli* of a functional multi-subunit ribulose bisphosphate carboxylase from a blue-green alga. Nature (Lond) 314:617–620

Gatenby AA, Lubben TH, Ahlquist P, Keegstra K (1988) Imported large subunits of ribulose bisphosphate carboxylase/oxygenase, but not imported β-ATP synthase subunits, are assembled into holoenzyme in isolated chloroplasts.EMBO J 7:1307–1314

Georgopoulos CP, Hendrix RW, Casjens SR, Kaiser AD (1973) Host participation in bacteriophage lambda head assembly. J Mol Biol 76:45–60

Goloubinoff P, Gatenby AA, Lorimer GH (1989) GroE heat-shock proteins promote assembly of foreign prokaryotic ribulose bisphosphate carboxylase oligomers in *Escherichia coli*. Nature (Lond) 337:44–47

Gutteridge S, Gatenby AA (1987) The molecular analysis of the assembly, structure and function of rubisco. In: Miflin BJ (ed) Oxford surveys of plant molecular and cell biology, Vol IV. Oxford Univ Press, Oxford, p 95

Hemmingsen SM, Ellis RJ (1986) Purification and properties of ribulosebisphosphate carboxylase large subunit binding protein. Plant Physiol 80:269–276

Hemmingsen SM, Woolford C, van der Vies SM et al. (1988) Homologous plant and bacterial proteins chaperone oligomeric protein assembly. Nature (Lond) 333:330–334

Hutchinson EG, Tichelaar W, Hofhaus G, Weiss H, Leonard KR (1989) Identification and electron microscopic analysis of a chaperonin oligomer from *Neurospora crassa* mitochondria. EMBO J 8:1485–1490

Jenkins AJ, March JB, Oliver IR, Masters M (1986) A DNA fragment containing the groE genes can suppress mutations in the *Escherichia coli* dnaA gene. Mol Gen Genet 202:446–454

Jindal S, Dudani AK, Singh B, Harley CB, Gupta RS (1989) Primary structure of a human mitochondrial protein homologous to the bacterial and plant chaperonins and to the 65-kilodalton mycobacterial antigen. Mol Cell Biol 9:2279–2283

Kochan J, Murialdo H (1983) Early intermediates in bacteriophage lambda prohead assembly. II. Identification of biologically active intermediates. Virol 131:100–115

Kochan J, Carrascosa JL, Murialdo H (1984) Bacteriophage lambda preconnectors: purification and structure. J Mol Biol 174:433–447

Lecker S, Lill R, Ziegelhoffer T et al. (1989) Three pure chaperone proteins of *Escherichia coli* -SecB, trigger factor and GroEL- form soluble complexes with precursor proteins in vitro. EMBO J 8:2703–2709

Lubben TH, Donaldson GK, Viitanen PV, Gatenby AA (1989) Several proteins imported into chloroplasts form stable complexes with the groEL-related chloroplast molecular chaperone. Plant Cell 1:1223–1230

McMullin TW, Hallberg RL (1988) A highly evolutionarily conserved mitochondrial protein is structurally related to the protein encoded by the *Escherichia coli* groEL gene. Mol Cell Biol 8:371–380

Musgrove JE, Johnson RA, Ellis RJ (1987) Dissociation of the ribulosebisphosphate-carboxylase large-subunit binding protein into dissimilar subunits. Eur J Biochem 163:529–534

Ostermann J, Horwich AL, Neupert W, Hartl F-U (1989) Protein folding in mitochondria requires complex formation with HSP60 and ATP hydrolysis. Nature (Lond) 342:125–129

Prasad TK, Hallberg RL (1989) Identification and metabolic characterization of the *Zea mays* mitochondrial homolog of the *Escherichia coli* groEL protein. Plant Mol Biol 12:609–618

Pushkin AV, Tsuprun VL, Solovjeva NA, Shubin VV, Evstigneeva ZG, Kretovich WL (1982) High molecular weight pea leaf protein similar to the groE protein of *Escherichia coli*. Biochim Biophys Acta 704:379–384

Reading DS, Hallberg RL, Myers AM (1989) Characterization of the yeast HSP60 gene coding for a mitochondrial assembly factor. Nature (Lond) 337:655–659

Roy H, Bloom M, Milos P, Monroe M (1982) Studies on the assembly of large subunits of ribulose bisphosphate carboxylase in isolated pea chloroplasts. J Cell Biol 94:20–27

Roy H, Hubbs A, Cannon S (1988) Stability and dissociation of the large subunit rubisco binding protein complex in vitro and in organello. Plant Physiol 86:50–53

Ruben SM, VanDenBrink-Webb SE, Rein DC, Meyer RR (1988) Suppression of the *Escherichia coli* ssb-1 mutation by an allele of groEL. Proc Natl Acad Sci USA 85:3767–3771

Schneider G, Lindqvist Y, Branden CI, Lorimer GH (1986) The three dimensional structure of ribulose-1,5-bisphosphate carboxylase/oxygenase from *Rhodospirillum rubrum* at 2.9A resolution. Embo J 5:3409–3415

Sternberg N (1973) Properties of a mutant of *Escherichia coli* defective in bacteriophage λ formation (groE). II. The propagation of phage λ. J Mol Biol 76:25–44

Straus DB, Walter WA, Gross CA (1988) *Escherichia coli* heat shock gene mutants are defective in proteolysis. Genes Dev 2:1851–1858

Tilly K, Murialdo H, Georgopolous C (1981) Identification of a second *Escherichia coli groE* gene whose product is necessary for bacteriophage morphogenesis. Proc Natl Acad Sci USA 78: 1629–1633

Van der Vies SM, Bradley D, Gatenby AA (1986) Assembly of cyanobacterial and higher plant ribulose bisphosphate carboxylase subunits into functional homologous and heterologous enzyme molecules in *Escherichia coli*. EMBO J 5:2439–2444

Van Dyk TK, Gatenby AA, LaRossa RA (1989) Genetic suppression demonstrates interaction of the GroE products with many proteins. Nature (Lond) 342:451–453

Waldinger D, Eckerskorn C, Lottspeich F, Cleve H (1988) Amino-acid sequence homology of a polymorphic cellular protein from human lymphocytes and the chaperonins from *Escherichia coli* (groEL) and chloroplasts (rubisco-binding protein). Biol Chem Hoppe-Seyler 369:1185–1189

Yu M-H, King J (1984) Single amino acid substitutions influencing the folding pathway of the phage P22 tail spike endorhamnosidase. Proc Natl Acad Sci USA 81:6584–6588

Zimmermann R, Sagstetter M, Lewis MJ, Pelham HRB (1988) Seventy-kilodalton heat shock proteins and an additional component from reticulocyte lysate stimulate import of M13 procoat protein into microsomes. EMBO J 7:2875–2880

Zweig M, Cummings D (1973) Cleavage of head and tail proteins during bacteriophage T5 assembly; Selective host involvement in the cleavage of a tail protein. J Mol Biol 80:505–518

CHAPTER 6

Mitochondrial Protein Import: Unfolding and Refolding of Precursor Proteins

Nikolaus Pfanner

Institut für Physiologische Chemie, Universität München, Goethestr. 33, D-8000 München 2, FRG

Introduction

More than 95% of mitochondrial proteins are encoded by nuclear genes, are synthesized as precursor proteins on cytosolic polysomes, and are eventually imported into one of the four mitochondrial subcompartments (outer membrane, intermembrane space, inner membrane, and matrix) (for review see Hurt and van Loon 1986; Attardi and Schatz 1988; Pfanner et al. 1988a; Hartl et al. 1989; Pfanner and Neupert 1989). The precursor proteins carry targeting information in amino-terminal extension sequences (termed presequences) or at various positions of the mature portion of the polypeptide chain. After recognition by receptor proteins on the mitochondrial surface, the precursor proteins become inserted into the outer membrane. Insertion into and translocation across the inner mitochondrial membrane occur at contact sites between both mitochondrial membranes and require the electrical potential $\Delta\Psi$ across the inner membrane. In the matrix, the presequences are proteolytically cleaved off. The proteins are then sorted to their functional destination and are assembled.

Proteins are not translocated in their mature (folded) conformation across the mitochondrial membranes. Precursor proteins in the cytosol have to be in a loosely folded ("unfolded") conformation that is compatible with translocation across the membranes. Later on the pathway, the proteins are (re)folded upon import into mitochondria. It was found that ATP and heat shock proteins are involved in both processes (Table 6.1). This review focuses on the mechanisms that modify and/or stabilize the conformation of mitochondrial precursor proteins. Furthermore, alternative mechanisms bypassing the requirements for ATP and heat shock proteins are described.

Translocation Competence of Precursor Proteins

Unfolding

A stable tertiary structure of a precursor protein interferes with its import into mitochondria. Eilers and Schatz (1986) employed a hybrid protein between a mitochondrial presequence (amino-terminal) and the cytosolic protein dihydro-

Stress Proteins
Schlesinger, Santoro, Garaci (Eds.)
© Springer-Verlag Berlin Heidelberg 1990

Table 6.1. Conformational changes of precursor proteins during import into mitochondria

Import step	Maintenance or conferring of a translocation-competent (loosely folded) conformation[a]	(Re)folding of imported proteins
Location	Cytosolic side	Matrix
Nucleoside triphosphate required	ATP or GTP (or other NTPs[b])	ATP
Components involved	HSP70, NEM-sensitive factor, possibly additional soluble and also membrane-bound components	HSP60, probably further components
Comments (alternative pathways)	Some precursor proteins can bypass the requirements for NTPs and/or cytosolic cofactors	Precursor of ADP/ATP carrier uses different pathway independent of ATP and HSP60

[a] Interaction of imported proteins with HSP60 in the matrix seems to be a prerequisite for retranslocation of proteins across the inner membrane; HSP60 and ATP may thus (at least in several cases) be involved in maintaining or conferring a translocation competent conformation on the matrix side.
[b] Abbreviations: NEM, N-ethylmaleimide; NTPs, nucleoside triphosphates; HSP70, heat shock protein(s) of 70 kDa; HSP60, heat shock protein of 60 kDa.

folate reductase (DHFR; carboxyl-terminal) to study conformational requirements of precursor proteins for import into mitochondria. Binding of a specific ligand (methotrexate) to the DHFR moiety induced a stable tertiary structure and thereby prevented uptake of the hybrid protein by mitochondria. Upon release of the ligand, the protein was imported and processed by mitochondria. Similar results were obtained by Chen and Douglas (1987a) with a different hybrid protein.

By arresting precursor proteins on their import pathway into mitochondria in a reversible manner (generation of translocation intermediates), it could be demonstrated that precursor proteins become (at least partially) unfolded during translocation across the mitochondrial membranes (Schleyer and Neupert 1985; Schwaiger et al. 1987; Pfanner et al. 1987a). Precursor proteins were trapped in contact sites between both mitochondrial membranes in a two-membrane spanning fashion: the amino-terminal presequence was cleaved off by processing peptidase in the mitochondrial matrix whilst carboxyl-terminal portions of the precursor protein were located on the cytosolic side of the outer membrane, as assessed by accessibility to externally added proteases. The distance across the two mitochondrial membranes is larger than the diameter of the mature folded protein, indicating that the precursor protein in contact sites was not completely folded. Recent studies showed that about 50 amino acid residues were sufficient to span the two mitochondrial membranes, suggesting a high degree of unfolding of the polypeptide chain (Rassow J, Hartl F-U, Guiard B, Pfanner N, Neupert W, in prep.).

Unfolding of domains of precursor proteins can occur on the mitochondrial surface while another portion of the precursor is already inserted into the mitochondrial protein import machinery. A hybrid protein between a relatively long amino-terminal portion derived from the authentic mitochondrial precursor protein cytochrome b2 and complete DHFR (carboxyl-terminal) was preincubated with methotrexate to induce a stable tertiary structure of the DHFR domain. Upon addition to mitochondria, the amino-terminal part became inserted into contact sites. The folded DHFR moiety remained on the cytosolic side of the membranes. After release of methotrexate, the DHFR part was also unfolded and the precursor became completely imported into mitochondria (Rassow et al. 1989).

Most studies on the unfolding of precursor proteins were performed in vitro with completely synthesized polypeptide chains, i.e., in a posttranslational import system. In the case of a cotranslational import reaction, a portion of the polypeptide chain is still buried in the ribosome while another part enters the mitochondrial membranes. This would imply that the polypeptide chain exhibits less tendency to fold into a stable tertiary structure; consequently, in cotranslational import there should be less requirement for unfolding of poly-peptide chains. It is well conceivable that a considerable fraction of import in vivo occurs during translation. Thus, the importance of unfolding processes for protein import in intact cells may be limited to a subset of precursor proteins.

HSP70 and Other Cytosolic Factors

Is unfolding of precursor proteins supported by membrane-bound components of the mitochondrial protein import apparatus or could cytosolic factors be involved? Several laboratories reported that cytosolic factors, possibly proteins, stimulated posttranslational import of proteins into isolated mitochondria (sum-marized in Pfanner and Neupert 1987; Murakami et al. 1988). Recently, yeast mutants deficient in a subset of 70 kDa stress proteins (heat shock proteins, HSP70) were found to be defective in mitochondrial protein import (Deshaies et al. 1988). Translocation of precursor proteins into isolated mitochondria was stimulated by addition of HSP70 (Murakami et al. 1988). As heat shock pro-teins have been ascribed a role in modulating conformation and interaction of proteins (Pelham 1986; Ellis and Hemmingsen 1989), it is assumed that they participate in conferring a translocation-competent conformation to precursor proteins or in maintaining such a conformation. Interestingly, HSP70 also stimul-ated protein translocation into the endoplasmic reticulum in vivo and in vitro (Deshaies et al. 1988; Chirico et al. 1988; Zimmermann et al. 1988) implying a possibly similar mechanism for an initial reaction of protein translocation across various membrane systems. Additional proteinaceous cytosolic factor(s) that were inactivated by the sulfhydryl reagent N-ethylmaleimide, were found to stimulate protein transport into mitochondria and ER (Murakami et al. 1988; Chirico et al. 1988; Randall and Shore 1989).

On the other hand, purified precursor proteins could be imported into mito-chondria without addition of cytosolic factors (Pfaller et al. 1985; Eilers and Schatz 1986). In one case, the precursor protein was artificially denatured ("un-folded") prior to its import into mitochondria (Pfaller et al. 1985). This supports the model that cytosolic factors are involved in conferring a translocation-conducive conformation and that "unfolded" precursor proteins may thus be independent of cytosolic factors (Pfanner et al. 1988b). In the second case, a hybrid protein between a mitochondrial presequence and DHFR carried a folded domain as assayed by the enzymatic activity of the DHFR moiety (Eilers and Schatz 1986). Apparently, some precursor proteins can employ different mechanisms without involvement of cytosolic factors to acquire translocation competence (also see below), although it cannot be excluded that small amounts of cytosolic factors were bound to the surface of the isolated mitochondria used in the import reaction. It remains to be elucidated whether heat shock proteins and other cytosolic factors are only required for import of some precursor proteins or if they play a more general role in import of a large number of precursor proteins.

A hybrid protein between a mitochondrial presequence and DHFR was partially unfolded on interaction with lipid vesicles. Endo and Schatz (1988) suggested that lipids of the mitochondrial outer membrane may similarly be involved in unfolding of precursor proteins. It remains to be determined if mitochondrial lipids directly interact with precursor proteins or if a possible role of lipids in protein import is more indirect via effects on embedded proteins of the import machinery.

Requirement for Nucleoside Triphosphates

Translocation of precursor proteins into mitochondria was found to require hydrolysis of nucleoside triphosphates (NTPs) (Pfanner and Neupert 1986; Eilers et al. 1987; Chen and Douglas 1987b). ATP may be the form of NTP required, but other NTPs such as GTP also stimulated protein import. Due to the presence of nucleoside phosphate kinases in the import systems, it was not possible so far to determine which NTP is actually necessary.

The present evidence suggests that NTPs are involved in conferring translocation competence to precursor proteins, in particular that NTPs participate in maintaining or generating a competent conformation: (1) Precursor proteins with identical targeting signals but different carboxyl-terminal portions were found to require rather different levels of NTPs for import (Pfanner et al. 1987b). (2) Loosely folded (or "unfolded") precursor proteins were imported into mitochondria also at (very) low levels of NTPs (Verner and Schatz 1987; Pfanner et al. 1988b). (3) The insertion of precursor proteins into the outer membrane that presumably depends on a high degree of unfolding of the polypeptides chain required the highest levels of NTPs as compared to other import steps such as binding to receptor sites (Pfanner et al. 1987b; Kleene et al. 1987). (4) When a precursor protein that was normally transported as tetramer in the cytosol was modified by deletion of the internal tetramer-forming sequence, the requirement of import for NTPs was strongly decreased (Chen and Douglas 1988). (5) Precursor proteins synthesized in rabbit reticulocyte lysates became more strongly folded upon depletion of NTPs (as assessed by their increased resistance towards digestion by low concentrations of proteases) (Pfanner et al. 1987b).

A model is conceivable where cytosolic cofactors, e.g., the ATP-dependent 70 kDa heat shock proteins, maintain or generate a loosely folded conformation of precursor proteins in a process requiring hydrolysis of ATP (Table 6.1). The cytosolic factors may bind and thereby stabilize a certain "unfolded" state of precursor proteins (Flynn et al. 1989); release of the precursor proteins from the cofactor(s), as a prerequisite for insertion of precursor proteins into the mitochondrial membranes, may then require hydrolysis of ATP. Moreover, the prevention of improper interactions between precursor proteins and other cytosolic components may involve ATP-dependent components (Pelham 1986; Ellis and Hemmingsen 1989).

It should be emphasized that exceptions to this general model are known. Two proteins with a relatively strongly folded domain, an authentic mitochondrial precursor protein (the precursor of F_0F_1-ATPase subunit 9) and a hybrid protein containing DHFR, were unfolded and imported at very low levels of NTPs (Pfanner et al. 1987b, 1990). In these cases, the degree of unfolding of the polypeptide chain that was necessary for translocation did not correlate with the amount of NTPs required. It is obvious that alternative or additional mechanisms have to exist to confer transport-competence to precursor proteins. This might involve ATP-independent cytosolic factors; however, membrane-bound components of the mitochondrial import machinery may also be of importance for the "unfolding" process. One could hypothesize that some precursor proteins can be transported in the cytosol without the aid of soluble factors and are directly bound, unfolded, and inserted by the membrane-bound import apparatus in an ATP-independent manner. Similarly. Eilers et al. (1988) reported that a hybrid protein containing DHFR was unfolded in the absence of NTPs; translocation across the mitochondrial membranes appeared to depend on NTPs. This latter step may involve the ATP-dependent interaction with the heat shock protein HSP60 in the mitochondrial matrix (Ostermann et al. 1989; see below).

(Re)folding of Imported Proteins in a Complex with HSP60

A heat shock protein of 60 kDa (HSP60), that is constitutively expressed and is localized in the mitochondrial matrix, plays an important role in (re)folding and assembly of proteins imported into the matrix. The first observation was made with a yeast mutant that was defective in functional HSP60, in a temperature-sensitive manner (Cheng et al. 1989). Precursor proteins could be imported into the mitochondrial matrix, but their assembly into multi-subunit complexes was impaired at the nonpermissive temperature. HSP60 and the homologous proteins of chloroplasts (Rubisco subunit binding protein) and *Escherichia coli* (GroEL) (Hemmingsen et al. 1988; Reading et al. 1989) were termed "chaperonins" since they appear to chaperone the assembly of oligomeric protein complexes (Ellis 1987; Ellis and Hemmingsen, 1989).

How does HSP60 mediate the assembly of protein complexes? The answer may come from the recent finding that HSP60 apparently possesses a more basic function, namely catalyzing the (re)folding of imported proteins independently of whether the proteins end up as monomers or oligomers (Ostermann et al. 1989). Modulation and/or stabilization of the conformation of imported proteins seems to be an essential function of HSP60. Most experimental results were obtained by use of a hybrid protein between a mitochondrial presequence and DHFR. The folding state of the DHFR moiety was monitored by its sensitivity towards digestion with proteinase K after lysis of mitochondria with detergent. Under control conditions, the imported protein was a stably folded monomer. When import was performed at low levels of ATP or in the presence of non-hydrolyzable ATP-analogs (the precursor protein had been denatured by urea to circumvent a possible ATP-dependent unfolding step on the cytosolic side), the hybrid protein accumulated as a loosely folded form in the matrix. This "unfolded" form was associated with HSP60. Readdition of ATP promoted (re)folding of this intermediate and release from HSP60, both in intact mitochondria and in mitochondrial extracts. When partially purified HSP60 with the bound hybrid protein was incubated in the presence of ATP, folding of the protein occurred on the surface of HSP60, but the protein was not released from HSP60 (Ostermann et al. 1989); additional factors that may be essential for release of the folded protein might have been lost during the purification procedure. This results demonstrated that folding of an imported protein can take place on the surface of HSP60 in an ATP-dependent process (Table 6.1).

Correct folding is most likely to be a prerequisite for assembly of proteins into multi-subunit complexes. If the assembly process *per se* is also mediated by HSP60 or other chaperonins is a question that has to be addressed in the future. Furthermore, retranslocation of some precursor proteins from the matrix into or across the inner membrane (Hartl et al. 1986, 1987) was defective in mitochondria from the HSP60-deficient yeast mutant (Cheng et al. 1989). HSP60 apparently is involved in conferring or stabilizing of a conformation necessary for reinsertion of proteins into the inner membrane (Ostermann et al. 1989).

Folding and assembly of proteins via HSP60 probably represents an essential step in sorting of most proteins destined for matrix, inner membrane, and intermembrane space. However, ADP/ATP carrier, the most abundant mitochondrial protein, is transported via contact sites to the inner membrane without depending on ATP for intramitochondrial sorting and assembly (Pfanner et al. 1987b). The assumption that import of ADP/ATP carrier does not require HSP60 was confirmed by the finding that its import and assembly was not impaired in mutant mitochondria deficient in HSP60 (Mahlke et al. 1990) (Table 6.1). Similar to the mechanisms of conferring translocation competence to precursor proteins in the cytosol, multiple pathways for intramitochondrial sorting and assembly appear to exist; a pathway via HSP60 including hydrolysis of ATP, and a quite distinct pathway (for ADP/ATP carrier and related proteins) that may involve lateral diffusion of the precursor proteins from contact sites to their functional destination in the inner membrane.

Perspectives

A competent conformation is an essential prerequisite for membrane translocation, sorting, and assembly of mitochondrial precursor proteins. Heat shock proteins, HSP70 in the cytosol and HSP60 in the matrix, seem to perform important functions in maintaining or conferring the appropriate conformation ["unfolded" or (re)folded] of the precursor proteins. Nucleoside triphosphates and probably further factors participate in these processes. Future research has to address the molecular details of interaction between heat shock proteins and precursor proteins and to identify additional components involved.

Of importance is the discovery of alternative pathways: some precursor proteins do not require HSP70 or NTPs for membrane translocation, whereas other precursor proteins bypass HSP60 and the requirement for ATP in the matrix. This emphasizes the complexity of mitochondrial protein import and cautions against minimal models.

Acknowledgments

The author thanks Drs. Walter Neupert and Franz-Ulrich Hartl for stimulating discussions and Drs. Walter Neupert and Rosemary Stuart for critical comments on the manuscript. Work from the author's laboratory was supported by the Deutsche Forschungsgemeinschaft (Sonderforschungsbereich 184, project B1).

References

Attardi G, Schatz (1988) Biogenesis of mitochondria. Ann Rev Cell Biol 4:289–333
Chen W-J, Douglas MG (1987a) The role of protein structure in the mitochondrial import pathway. Unfolding of mitochondrially bound precursors is required for membrane translocation. J Biol Chem 262:15605–15609
Chen W-J, Douglas MG (1987b) Phosphodiester bond cleavage outside mitochondria is required for the completion of protein import into the mitochondrial matrix. Cell 49:651–658
Chen W-J, Douglas MG (1988) An F_1-ATPase β-subunit precursor lacking an internal tetramer-forming domain is imported into mitochondria in the absence of ATP. J Biol Chem 263:4997–5000
Cheng MY, Hartl F-U, Martin J et al. (1989) Mitochondrial heat-shock protein hsp60 is essential for assembly of proteins imported into yeast mitochondria. Nature (Lond) 337:620–625
Chirico WJ, Waters MG, Blobel G (1988) 70K heat shock related proteins stimulate protein translocation into microsomes. Nature (Lond) 332:805–810
Deshaies RJ, Koch BD, Werner-Washburne M, Craig EA, Schekman R (1988) A subfamily of stress proteins facilitates translocation of secretory and mitochondrial precursor polypeptides. Nature (Lond) 332:800–805
Eilers M, Schatz G (1986) Binding of a specific ligand inhibits import of a purified precursor protein into mitochondria. Nature (Lond) 322:228–232
Eilers M, Oppliger W, Schatz G (1987) Both ATP and an energized inner membrane are required to import a purified precursor protein into mitochondria. EMBO J 6:1073–1077
Eilers M, Hwang S, Schatz G (1988) Unfolding and refolding of a purified precursor protein during import into isolated mitochondria. EMBO J 7:1139–1145
Ellis J (1987) Proteins as molecular chaperones. Nature (Lond) 328:378–379

Ellis RJ, Hemmingsen SM (1989) Molecular chaperones: proteins essential for the biogenesis of some macromolecular structures. Trends Biochem Sci 14:339–342

Endo T, Schatz G (1988) Latent membrane perturbation activity of a mitochondrial precursor protein is exposed by unfolding. EMBO J 7:1153–1158

Flynn GC, Chappell TG, Rothman JE (1989) Peptide binding and release by proteins implicated as catalysts of protein assembly. Science 245:385–390

Hartl F-U, Schmidt B, Weiss H, Wachter E, Neupert W (1986) Transport into mitochondria and intramitochondrial sorting of the Fe/S protein of ubiquinol cytochrome c reductase. Cell 47: 939–951

Hartl F-U, Ostermann J, Guiard B, Neupert W (1987) Successive translocation into and out of the mitochondrial matrix: targeting of proteins to the intermembrane space by a bipartite signal peptide. Cell 51:1027–1037

Hartl F-U, Pfanner N, Nicholson DW, Neupert W (1989) Mitochondrial protein import. Biochim Biophys Acta 988:1–45

Hemmingsen SM, Woolford C, van der Vies SM et al. (1988) Homologous plant and bacterial proteins chaperone oligomeric protein assembly. Nature (Lond) 333:330–334

Hurt EC, van Loon APGM (1986) How proteins find mitochondria and intramitochondrial compartments. Trends Biochem Sci 11:204–207

Kleene R, Pfanner N, Pfaller R et al. (1987) Mitochondrial porin of Neurospora crassa: cDNA cloning, in vitro expression and import into mitochondria. EMBO J 6:2627–2633

Mahlke K, Pfanner N, Martin J, Horwich AL, Hartl F-U, Neupert W (1990) Sorting pathways of mitochondrial inner membrane proteins. Eur J Biochem, in press

Murakami H, Pain D, Blobel G (1988) 70-kD heat shock-related protein is one of at least two distinct cytosolic factors stimulating protein import into mitochondria. J Cell Biol 107:2051–2057

Ostermann J, Horwich AL, Neupert W, Hartl F-U (1989) Protein folding in mitochondria requires complex formation with hsp60 and ATP hydrolysis. Nature (Lond) 341:125–130

Pelham HRB (1986) Speculations on the functions of the major heat shock and glucose-regulated proteins. Cell 46:959–961

Pfaller R, Freitag H, Harmey MA, Benz R, Neupert W (1985) A water-soluble form of porin from the mitochondrial outer membrane of Neurospora crassa: properties and relationship to the biosynthetic precursor form. J Biol Chem 260:8188–8193

Pfanner N, Neupert W (1986) Transport of F_1-ATPase subunit β into mitochondria depends on both a membrane potential and nucleoside triphosphates. FEBS Lett 209:152–156

Pfanner N, Neupert W (1987) Biogenesis of mitochondrial energy transducing complexes. Curr Top Bioenerg 15:177–219

Pfanner N, Neupert W (1989) Transport of proteins into mitochondria. Curr Opin Cell Biol 1: 624–629

Pfanner N, Hartl F-U, Guiard B, Neupert W (1987a) Mitochondrial precursor proteins are imported through a hydrophilic membrane environment. Eur J Biochem 169:289–293

Pfanner N, Tropschug M, Neupert W (1987b) Mitochondrial protein import: nucleoside triphosphates are involved in conferring import-competence to precursors. Cell 49:815–823

Pfanner N, Hartl F-U, Neupert W (1988a) Import of proteins into mitochondria: a multi-step process. Eur J Biochem 175:205–212

Pfanner N, Pfaller R, Kleene R, Ito M, Tropschug M, Neupert W (1988b) Role of ATP in mitochondrial protein import: conformational alteration of a precursor protein can substitute for ATP requirement. J Biol Chem 263:4049–4051

Pfanner N, Rassow J, Guiard B, Söllner T, Hartl F-U, Neupert W (1990) Energy requirements for unfolding and membrane translocation of precursor proteins during import into mitochondria. J Biol Chem, in press

Randall SK, Shore GC (1989) Import of a mutant mitochondrial precursor fails to respond to stimulation by a cytosolic factor. FEBS Lett 250:561–564

Rassow J, Guiard B, Wienhues U, Herzog V, Hartl F-U, Neupert W (1989) Translocation arrest by reversible folding of a precursor protein imported into mitochondria. A means to quantitate translocation contact sites. J Cell Biol 109:1421–1428

Reading DS, Hallberg RL, Myers AM (1989) Characterization of the yeast HSP60 gene coding for a mitochondrial assembly factor. Nature (Lond) 337:655–659

Schleyer M, Neupert W (1985) Transport of proteins into mitochondria: translocational intermediates spanning contact sites between outer and inner membranes. Cell 43:339–350

Schwaiger M, Herzog V, Neupert W (1987) Characterization of translocation contact sites involved in the import of mitochondrial proteins. J Cell Biol 105:235–246

Verner K, Schatz G (1987) Import of an incompletely folded precursor protein into isolated mitochondria requires an energized inner membrane, but no added ATP. EMBO J 6:2449–2456

Zimmermann R, Sagstetter M, Lewis MJ, Pelham HRB (1988) Seventy-kilodalton heat shock proteins and additional component from reticulocyte lysate stimulate import of M13 procoat protein into microsomes. EMBO J 7:2875–2880

CHAPTER 7

The Ubiquitin System and the Heat Shock Response

Milton J. Schlesinger

Department of Molecular Microbiology, Washington University School of Medicine,
St. Louis, MO 63110, USA

A cell subjected to an increase in temperature as little as 10% above the physiol-
ogically normal range suffers damage that can permanently affect its growth
and function. There are, however, a number of activities that protect the cell
from a heat shock as well as other kinds of environmental stress. Research over
the past 15 years has clearly established that one of these protective mechanisms
is the induction of heat shock proteins. Chapters in this volume provide important clues as to how the synthesis of the major universal stress protein HSP70 is
regulated and how it functions as a "chaperone" to form complexes with proteins that misfold or unfold during stress and thus "rescue" these proteins
from irreversible damage and degradation. Two recent reviews (Pelham 1989;
Rothman 1989) elaborate further on functions of several proteins closely related
in structure to the major heat shock proteins. But not all stress-damaged proteins can be rescued and it is now clear that stressed cells also activate several
components of a proteolytic degradation system that normally functions to "turn
over" cytoplasmic and nuclear proteins. This enzymatic pathway is found in all
eukaryotic cells and employs the small polypeptide called ubiquitin to mark a
protein for proteolysis. Many of the enzymes utilized in ubiquitin-dependent
proteolysis have been isolated and examined in detail (Hershko 1988), and
recent studies show that several of these are induced by heat shock and other
stressors. In this chapter I review these developments and the relationship of the
ubiquitin degradation system to the cell's stress response program.

Ubiquitin is a 76 amino acid polypeptide found in large amounts (10^6 molecules per cell) in virtually all eukaryotic cells. The three-dimensional atomic
structure, based on X-ray crystallographic data and two-dimensional ^{1}H-NMR
(Vijay-kumar et al. 1987; DiStefano and Wand 1987), shows ubiquitin to be a
compact globular protein. Four amino acids, leu.arg.gly.gly, at the carboxy
terminus extend outward from the structure and the carboxyl-terminal glycine
can form an isopeptide bond to an epsilon-amino group of a lysine in a "target"
protein. Ubiquitin is one of the most highly conserved proteins known: only
three amino acids in the sequence differ from yeast to human ubiquitin without
any change in tertiary structure.

There are two kinds of genes encoding ubiquitin sequences and both have
unusual structures in their reading frames (Schlesinger and Bond 1987). One
type — the polyubiquitin gene — has multiple sequences of ubiquitin arranged

Stress Proteins
Schlesinger, Santoro, Garaci (Eds.)
© Springer-Verlag Berlin Heidelberg 1990

contiguously without stop or start signals. The number of repeated ubiquitins ranges from three in a human gene to greater than 50 in the trypanosome (Kirchoff et al. 1988; Swindle et al. 1988). Several polyubiquitin genes occur in a cell's genome and at least one contains a heat shock promoter sequence in the 5′ non-transcribed region (Bond and Schlesinger 1985). The second type of gene contains ubiquitin coding sequences fused at the carboxyl terminus in-frame to sequences coding for ribosomal proteins (Ozkaynak et al. 1987). In one of these latter-type genes, there is a sequence of 52 amino acids coding for a protein of the 60S subunit and in another a sequence of 76–80 amino acids coding for a protein of the 40S subunit of the ribosome (Finley et al. 1989; Redman and Rechsteiner 1989; Müller-Taubenberger et al. 1989). In *S.cerevisiae* there are two copies of the gene encoding the large ribosomal subunit protein. Both kinds of these fusion genes are expressed during cell growth (Finley et al. 1989).

Ubiquitin interacts with the three major macromolecules of the cell — DNA, RNA, and protein — in the manner depicted in Fig. 7.1. The first reported role for ubiquitin, which was initially isolated as a thymocyte growth factor (Goldstein et al. 1975), was in DNA structure where it was found linked covalently to the histones H2A and H2B (Goldknopf and Busch 1977). Both histones have a single ubiquitin bound in isopeptide linkage to a specific lysine near the carboxyl terminus of the histone sequence. The precise function of the ubiquitinated histones is unknown but a variety of the data published over the past 10 years indicates that ubiquitination is essential for DNA to be "active" in chromatin. Condensed chromatin is devoid of ubiquitinated histones (Matsui et al. 1979; Wu et al. 1981; Mueller et al. 1985) and the latter are enriched in DNA fractions undergoing transcription and replication (Nickel et al. 1989). One of the

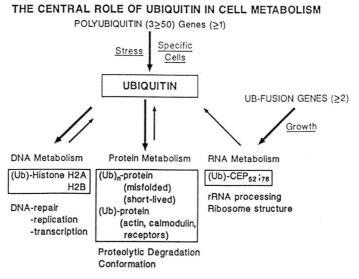

Fig. 7.1. Central role of ubiquitin in cell macromolecular metabolism

genes involved in DNA repair, noted as RAD6 (Prakash 1981), of *S.cerevisiae* encodes an enzyme required to ubiquitinate H2B (Jentsch et al. 1987; Sung et al. 1988). Another *S.cerevisiae* gene, cdc34, encodes a protein needed for cell division and also ubiquitinates H2B (Goebl et al. 1988). Histone ubiquitination is also blocked in a temperature-sensitive mutant of a murine cell line that is defective in cell division. This mutation resides in the E1 ubiquitin activation enzyme (see below, Finley et al. 1984). All of these data show that histone ubiquitination is an essential step in normal DNA metabolism. In heat-shocked cells, ubiquitinated histones disappear rapidly (Glover 1982; Bond et al. 1988). It is likely that this activity is associated with the condensation of chromatin detected in stressed mammalian cells (Welch and Suhan 1985) and to the rapid decrease in DNA synthesis found after stress (Schlesinger et al. 1989). A ubiquitin hydrolase has been described that removes ubiquitin from histones (Andersen et al. 1981; Matsui et al. 1982). This enzyme may be activated by stress and the released ubiquitin could exit the nucleus and participate in the ubiquitin-dependent proteolytic degradation pathway (see below).

Ubiquitin enters the RNA metabolic pathway by virtue of its covalent association with two ribosomal proteins described earlier. The ubiquitin moiety of the fusion protein is postulated to stabilize the ribosomal protein during its synthesis (Finley et al. 1989) and transport it to the site of assembly in the nucleolous. Ribosome assembly is quite sensitive to stress and ribosomal RNA processing slows after heat shock with accumulation of the large precursor RNAs (Schlesinger et al. 1989). In addition, shortly after a stress the major heat shock protein, HSP70, moves to the nucleolous where it binds tightly to proteins in this organelle (Munro and Pelham 1984).

The ubiquitin-fusion proteins are the major de novo source of free ubiquitin in exponentially growing *S.cerevisiae* (Finley et al. 1987); however, this source of ubiquitin becomes limiting very shortly after a stress and additional ubiquitin

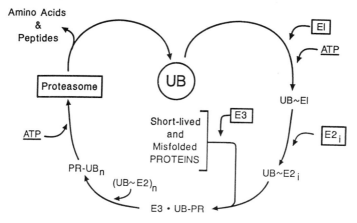

Fig. 7.2. Enzymatic pathway for ubiquitin-dependent proteolytic degradation. E1, E2i isoforms, and E3 are decribed in the text. *UB-PR* refers to a mono-ubiquitinated protein; *PR-UBn* refers to a polyubiquitinated protein. The proteasome is described in the text

is obtained by activation of a polyubiquitin gene. During the stress, transcription of the ubiquitin-fusion genes shuts off thereby limiting the amount of two ribosomal proteins. Ribosome assembly stops and this could account for the decreased level of ribosomal RNA processing noted above.

The role of ubiquitin in protein metabolism has received the major attention of investigators studying this protein and the biochemical pathway whereby ubiquitin acts as the cofactor for cytoplasmic ATP-dependent protein degradation is well established (Fig. 7.2). Three kinds of enzymes — E1, E2, and E3 — participate in the activation and transfer of ubiquitin to a target protein. Six isoforms of E2 and two of E3 have been isolated thus far. The E1 activates ubiquitin in an ATP-dependent reaction and transfers it to one of the E2s. E3 recognizes a target protein and transfers ubiquitin from a specific E2 to the protein. What determines whether a protein will be a target for ubiquitination has been under intense study and has led recently to the formulation of the "N-end" rule (Gonda et al. 1989). This rule defines the amino-terminus of a protein as a major feature in its susceptibility to ubiquitin-dependent degradation. Proteins with charged amino acids or bulky hydrophobic residues at their amino terminus are most susceptible to ubiquitination. However, genetic studies in *S.cerevisiae* indicate that the "N-end" rule is probably not the major determinant for protein turnover (A.Varshavsky pers. commun. 1989). Reiss et al.

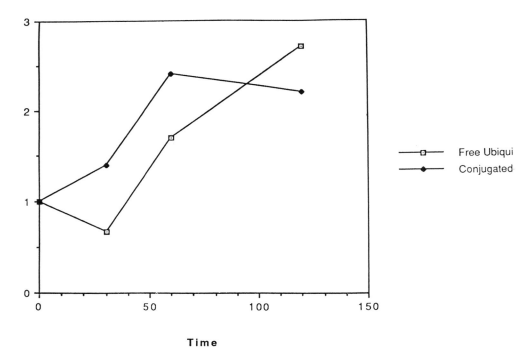

Fig. 7.3. Changes in ubiquitin conjugates and free ubiquitin after heat shock in chicken embryo fibroblasts. Cells were heat-shocked to 45 °C for 60 min and returned to 38 °C for 60 min. Samples were analyzed at 0, 30, 60, and 120 min. (Details in Bond et al. 1988)

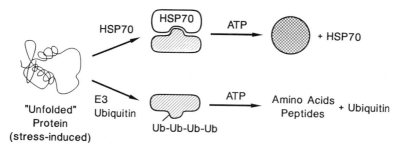

Fig. 7.4. Alternative fates for an unfolded protein in a stressed cell. The structure with *slanted lines* is the unfolded protein complexed with either HSP70 or polyubiquitin. The structure with *hatched lines* is the refolded protein

(1988) also showed that these same kinds of amino-terminal amino acids were important determinants in ubiquitin targeting but, in addition, found that there were other features of a protein besides its amino terminus that were recognized by an E3 protein. Based on these results, Hershko proposed that a protein's "body" could be an important factor in its stability; thus, a protein that fails to fold properly or partially unfolds might be recognized by this isoform of E3 and become ubiquitinated (Heller and Hershko 1990). These kinds of aberrant proteins are formed in cells stressed by heat shock, organic solvents, and oxidants and levels of ubiquitin protein conjugates increase almost three-fold in a stressed cell (Fig. 7.3) (Parag et al. 1987). In addition, levels of the mRNAs for at least two of the E2 type proteins from *S.cerevisiae* increase severalfold after stress. Mutations in the genes for these enzymes block formation of high molecular weight conjugates and make the cell exceedingly sensitive to canavanine, an analog which substitutes for arginine and forms abnormal proteins (Seufert and Jentsch 1990). Synthesis of the isoform of E3 that recognizes unfolded proteins might also be activated by stress and it could compete with HSP70 in binding to unfolded polypeptides (Fig. 7.4). Formation of a HSP70-complex would rescue the protein whereas that with E3 would lead to its degradation. The "rescue" pathway would prevail in cells containing abundant amounts of HSP70 proteins prior to a severe stress. This postulated competition between ubiquitin E3 and HSP70 provides one molecular mechanism for the phenomenon of thermotolerance (see Chap. 9 this vol).

    The continued formation of unfolded proteins in the stressed cell increases the demand for additional ubiquitin molecules that are supplied by activation of a polyubiquitin gene (see above and Fig. 7.1). Translation of the mRNAs transcribed from these genes produces a polyubiquitin that is proteolytically processed to the individual ubiquitin molecules. Synthesis or activation of this putative polyubiquitin protease also might be stress-regulated. The unfolded proteins recognized by the ubiquitin E3 enzyme are tightly bound to the latter (Reiss et al. 1988) and become multiubiquitinated as more E2-ubiquitins are shuttled onto the E3-protein complex (see Fig. 7.2). Recent evidence indicates that a "tree" of ubiquitin forms with one ubiquitin bound to the protein and the others bound to lys48 of ubiquitin (Chau et al. 1989). The poly-ubiquitinated protein is degraded by a multicomponent organelle of high molecular weight with a sedimentation coefficient of 26S (Ganoth et al. 1988). A 19–20S form of this organelle contains the proteases but is not ubiquitin-dependent in peptide degradation — it has been termed the proteasome (Matthews et al. 1989, Fig. 7.2). Conversion of the 20S particle to the ubiquitin-dependent form requires ATP (Eytan et al. 1989). Some 13 distinct subunits have been identified as components of the proteasome and several of these polypeptides from liver, erythrocyte, *Drosophila*, and yeast have been cloned and sequenced (Tanaka, Slaughter, Ploegh, and Heinemann pers. commun. 1989). The subunits range in size from 20 to 35 kDa and are related in sequence. Proteolytic activities are blocked partially by trypsin and chymotrypsin inhibitors, but none of the subunit sequences examined thus far shows catalytic sites common to the four known types of proteases. The ATP-dependent proteasome (the 19–20S species) complexed with additional proteins, one of which has a putative ubiquitin recognition site, is clearly a critical organelle for eliminating the high molecular weight ubiquitin-protein complexes that accumulate in stressed cells. Because of its complex structure, it is unlikely that the amount of this organelle would change substantially during a stress; thus the final elimination of misfolded proteins would be rate-limited by activity of the proteasome complex. In cases where there is prolonged stress or extensive cell damage, high molecular weight ubiquitin conjugates might aggregate in or around cells. Such aggregates have been found in many aberrant structures associated with diseases of the central nervous system such as Alzheimer and Parkinson disease (Lowe et al. 1988).

A summary of those parts of the ubiquitin system affected by and important to the stressed cell is presented in Table 7.1. It shows that many genes of the ubiquitin system encode heat shock proteins. Thus, ubiquitin and enzymes involved in its metabolism are an integral part of the evolutionarily conserved cellular mechanism that protects the cell from stress damage and allows it to resume a normal life style after removal of the stress.

Table 7.1. Response of ubiquitin components to heat shock

Component	Response
Polyubiquitin gene	mRNA levels up 4–8-fold
Ubiquitin-fusion genes	mRNA levels decrease
Ubiquitin-activating enzyme (E1)	mRNA levels unchanged
Ubiquitin-transfer enzymes (E2)	mRNA levels for two isoforms increase
Ubiquitin-targeting enzyme (E3)	mRNA levels unchanged for the isoform obeying the N-end rule
Ubiquitinated histone H2A,B	Levels decrease
Ubiquitinated proteins (high mol.wt.conjugates)	Levels increase
Ubiquitin hydrolases (1) Polyubiquitin (2) Ubiquitin-histone (3) Ubiquitin-fusion protein (4) Ubiquitin-peptide	All are *postulated* to increase in activity but not as a result of increased gene activation or mRNA translation

References

Andersen MW, Balla NR, Goldknopf IL, Busch H (1981) Protein A24 lyase activity in nucleoli of thioacetamide-treated rat liver releases histone H2A and ubiquitin from conjugate protein A24. Biochemistry 20:1100–1104

Bond U, Schlesinger MJ (1985) Ubiquitin is a heat shock protein in chicken embryo fibroblasts. Mol Cell Biol 5:949–956

Bond U, Agell N, Haas AL, Redman K, Schlesinger MJ (1988) Ubiquitin in stressed chicken embryo fibroblasts. J Biol Chem 263:2384–2388

Chau V, Tobias JW, Bachmair A et al. (1989) A multiubiquitin chain is confined to specific lysine in a targeted short-lived protein. Science 243:1576–1583

DiStefano DL, Wand AJ (1987) Two-dimensional ^{1}H NMR study of human ubiquitin: a main chain directed assignment and structure analysis. Biochemistry 26:7272–7281

Eytan E, Ganoth D, Armon T, Hershko A (1989) ATP-dependent incorporation of 20S protease into the 26S complex that degrades protein conjugated to ubiquitin. Proc Natl Acad Sci USA 86:7751–7755

Finley D, Ciechanover A, Varshavsky A (1984) Thermolability of ubiquitin-activating enzyme from the mammalian cell cycle mutant ts85. Cell 37:43–55

Finley D, Ozkaynak E, Varshavsky A (1987) The yeast polyubiquitin gene is essential for resistance to high temperatures, starvation, and other stresses. Cell 48:1035–1046

Finley D, Bartel B, Varshavsky A (1989) The tails of ubiquitin precursors are ribosomal proteins whose fusion to ubiquitin facilitates ribosome biogenesis. Nature (Lond) 338:394–401

Ganoth D, Leshinsky E, Eytan E, Hershko A (1988) A multicomponent system that degrades proteins conjugated to ubiquitin. J Biol Chem 263:12412–12419

Glover CVC (1982) Heat shock effects on protein phosphorylation in *Drosophila*. In: Schlesinger MJ, Ashburner M, Tissieres A (eds) Heat shock from bacteria to man. Cold Spring Harbor Lab, Cold Spring Harbor, p 227

Goebl MG, Yochem J, Jentsch S, McGrath J, Varshavsky A, Beyers B (1988) The yeast cell cycle gene CDC34 encodes a ubiquitin-conjugating enzyme. Science 241:1331–1335

Goldknopf IL, Busch H (1977) Isopeptide linkage between nonhistone and histone 2A polypeptides of chromosomal conjugate-protein A24. Proc Natl Acad Sci USA 74:864–868

Goldstein G, Scheid M, Hammerling U, Boyce EA, Schlesinger DH, Niall HD (1975) Isolation of a polypeptide that has lymphocyte-differentiating properties and is probably represented universally in living cells. Proc Natl Acad Sci USA 72:11–15

Gonda DK, Bachmair A, Wunning I, Tobias JW, Lane WS, Varshavsky A (1989) Universality and structure of the N-end rule. J Biol Chem 264:16700–16712

Heller H, Hershko A (1990) A ubiquitin-protein ligase specific for type III protein substrates. J Biol Chem 265:6532–6535

Hershko A (1988) Ubiquitin-mediated protein degradation. J Biol Chem 263:15237–15240

Jentsch S, McGrath JP, Varshavsky A (1987) The yeast DNA repair gene RAD6 encodes a ubiquitin-conjugating enzyme. Nature (Lond) 329:131–134

Kirchoff LV, Kwang SK, Engman DM, Donelson JE (1988) Ubiquitin genes in Trypanosomatidae. J Biol Chem 263:12698–12704

Lowe J, Blanchard A, Morrel K et al. (1988) Ubiquitin is a common factor in intermediate filament inclusion bodies of diverse type in man, including those of Parkinson's disease, Pick disease and Alzheimer disease, as well as Rosenthal fibres in cerebellar astrocytomas, cytoplasmic bodies in muscle, and Mallory bodies in alcoholic liver disease. J Pathol 155:9–15

Matsui S-I, Seon BK, Sandberg AA (1979) Disappearance of a structural chromatin protein A24 in mitosis: Implications for molecular basis of chromatin condensation. Proc Natl Acad Sci USA 76:6386–6390

Matsui S-I, Sandberg AA, Negoro S, Seon BK, Goldstein G (1982) Isopeptidase: a novel eukaryotic enzyme that cleaves isopeptide bonds. Proc Natl Acad Sci USA 79:1535–1539

Matthews W, Tanaka K, Driscoll J, Ichikara A, Goldberg AL (1989) Involvment of the proteasome in various degradative processes in mammalian cells. Proc Natl Acad Sci USA 81:2597–2601

Mueller RD, Yasada H, Hatch CL, Bonner WM, Bradbury EM (1985) Identification of ubiquitinated H2A and H2B in *Physarum polychephalum*: disappearance of these proteins at metaphase and reappearance at anaphase. J Biol Chem 260:5147–5153

Muller-Taubenberger A, Graack HR, Grohmann L, Schleicher M, Gerisch G (1989) An extended ubiquitin of *Dictyostelium* is located in the small ribosomal subunit. J Biol Chem 264:5319–5322

Munro S, Pelham HRB (1984) Use of peptide tagging to detect proteins expressed from cloned genes: deletion mapping functional domains of Drosophila HSP70. EMBO J 3:3087–3093

Nickel BE, Allis CD, Davie JR (1989) Ubiquitinated histone H2B is preferentially located in transcriptionally active chromatin. Biochemistry 28:958–963

Ozkaynak E, Finley D, Solomon MJ, Varshavsky A (1987) The yeast ubiquitin genes: a family of natural gene fusions. EMBO J 6:1429–1439

Parag HS, Raboy B, Kulka RG (1987) Effect of heat shock on protein degradation in mammalian cells: involvment of the ubiquitin system. EMBO J 6:55–61

Pelham HRB (1989) Heat shock and the sorting of luminal ER proteins. EMBO J 8:3171–3176

Prakash L (1981) Characterization of postreplication repair in Saccharomyces cerevisiae and effects of rad6, rad18, rev3 and rad5 mutations. Mol Gen Genet 184:471–478

Redman KL, Rechsteiner M (1989) Identification of the long ubiquitin extension as a ribosomal protein S27A. Nature (Lond) 338:438–440

Reiss Y, Kaim D, Hershko A (1988) Specificity of binding of NH_2-terminal residue of proteins to ubiquitin-protein ligase. J Biol Chem 263:2693–2698

Rothman JE (1989) Polypeptide chain binding proteins: catalysts of protein folding and related processes in the cell. Cell 59:591–601

Schlesinger MJ, Bond U (1987) Ubiquitin genes. Oxf Surv Euk Genes 4:77–91

Schlesinger MJ, Collier NC, Agell N, Bond U (1989) Molecular events in avian cells stressed by heat shock and arsenite. In: Pardue ML, Feramisco JR, Lindquist S (eds) Stress-induced proteins. Alan R Liss, New York, p 137

Seufert W, Jentsch S (1990) Ubiquitin-conjugating enzymes UBC4 and UBC5 mediate selective protein degradation, a central growth function essential during heat shock. EMBO J 9:543–550

Sung, P, Prakash S, Prakash L (1988) The RAD6 protein of Saccharomyces cerevisiae polyubiquitinates histones, and its acidic domain mediates this activity. Genes Dev 2:1476–1485

Swindle J, Ajioka J, Eisen H et al. (1988) the genomic organization and transcription of the ubiquitin genes of Trypanosoma cruzi. EMBO J 7:1121–1127

Vijay-kumar S, Bugg CE, Cook WJ (1987) Structure of ubiquitin refined at 1.8 Å resolution. J Mol Biol 194:531–544

Welch WJ, Suhan JP (1985) Morphological study of the mammalian stress response: characterization of changes in cytoplasmic organelles, cytoskeleton, and nucleoli, and the appearance of intranuclear actin filaments in rat fibroblasts after heat shock treatment. J Cell Biol 101:1198–1211

Wu RS, Kohn KW, Bonner WM (1981) Metabolism of ubiquitinated histones. J Biol Chem 256:5916–5920

CHAPTER 8

Protein Denaturation During Heat Shock and Related Stress

Olivier Bensaude, Moise Pinto, Marie-Françoise Dubois, Nguyen Van Trung and Michel Morange

Groupe de Biologie Moléculaire du Stress, Institut Pasteur, 25 rue du Dr. Roux, 75724 Paris Cedex 15, France

In this chapter, we discuss two questions raised by heat shock studies: what are the effects of stress on cells, and is the role of heat shock proteins to prevent as well as cure cell damage?

Cell Damage During Heat Shock

Heat shock damages cells and, upon increasingly severe stress conditions, leads to cell death. Among the various cellular functions and structures impaired by heat are rRNA synthesis (Black and Subject 1989; Parker and Bond 1989), premRNA splicing (Yost and Lindquist 1986), protein synthesis, DNA replication (Mivecchi and Dewey 1985), and cytoskeletal structure (Welch and Suhan 1986). These might be due to the inactivation of some crucial enzymatic steps, and, in fact several enzymes have been shown to be inactivated during hyperthermia. These include DNA polymerases (Spiro et al. 1982), poly(ADP-ribose) synthetase (Nolan and Kidwell 1982), Na^+/K^+-ATPases (Anderson and Hahn 1985), Ca^{2+}-ATPase (Cheng et al. 1987), and NADPH oxidase (Maridonneau-Parini et al. 1988). Moreover, binding of insulin (Calderwood and Hahn 1983), epidermal growth factor (Magun and Fennle 1981), and steroids (Wolffe et al. 1984) to their respective receptors is inhibited by heat shock.

Nature of Abnormal Proteins

Loss of protein function might be a consequence of some critical modifications occurring during stress. Subtle modifications have been described: the phosphorylation state of the eukaryotic initiation factors eIF-2 and eIF-4 and of the ribosomal protein S6 are altered by heat and these could contribute to the reversible inactivation of protein synthesis (Kennedy et al. 1984; Duncan et al. 1987). Changes in the phosphorylation, methylation, ubiquitination, and acetylation of histones might be responsible for the observed alterations in

Stress Proteins
Schlesinger, Santoro, Garaci (Eds.)
© Springer-Verlag Berlin Heidelberg 1990

chromatin structure (Glover et al. 1981; Camato and Tanguay 1982; Arrigo 1983).

Loss of protein function might also be a consequence of nonspecific denaturation. Indeed, abnormal proteins were expected to appear during a stress (Hightower 1980), but their origin is still unclear. Several hypotheses have been proposed: heat per se might act as a denaturing agent, or alterations in cellular metabolism might cause deleterious modifications of some proteins. For example, heat shock could generate oxygen radicals which would damage proteins (Ropp et al. 1983; Burdon et al. 1987; Drummond and Steinhardt 1987).

Fate of "Abnormal" Proteins: Degradation or Aggregation

Under standard culture conditions, abnormal proteins can be degraded by the La protease in *E. Coli*, whereas in eukaryotic cells, abnormal proteins are targets for ubiquitin-dependent degradation (Finley et al. 1984). During heat shock, ubiquitinated proteins increase (Carlson et al. 1987; Bond et al. 1988) and protein degradation increases after heat shock through the ubiquitin pathway (Parag et al. 1987; Westwood and Steinhardt 1989). Interestingly, ubiquitin in eukaryotic cells and the La protease in *E. Coli* are both heat shock proteins (Goff and Goldberg 1985; Bond and Schlesinger 1985). Alternatively, abnormal proteins might form insoluble aggregates. Biotechnologists are well aware that overproduction of a protein within a transgenic cell often leads to the formation of inclusion bodies which are almost pure aggregates of that protein (Kane and Hartley 1988). These aggregates result from an incorrect folding of the polypeptide chains (Pain 1987).

Insolubilization of Foreign Reporter Proteins During Stress

To study the effect of stress on a protein's native structure, we introduced foreign "reporter" enzymes into cells (Nguyen et al. 1989). Enzymes were considered to be attractive model proteins since they allow for the simultaneous analysis of biological activity and physiochemical properties such as solubility. Mouse fibroblasts were transfected with genes coding for either firefly luciferase or bacterial β-galactosidase. When cells were grown at 37 °C and lysed in buffers containing nonionic detergents, both enzymes were cytoplasmic and soluble. After a heat shock, enzyme activities decreased in the postmitochondrial supernatants (10,000 g). Total lysates plus supernatant and pellet fractions of insoluble material were analyzed by Western blot. After heat shock, the reporter proteins were found in the pellets in amounts proportional to the loss of their activity. No changes in electrophoretic mobilities of the reporter proteins were detected in SDS/PAGE, and immunochemical staining showed a diffuse pattern for both proteins in the cytoplasm with little change in their distribution after stress.

Insolubilization of Cellular Proteins During Stress

A dramatic loss in solubility of the nuclear oncogenes and protooncogenes of the myc, myb, and p53 families occurs in heat-shocked cells (Evan and Hancock 1985; Littlewood et al. 1987; Lüscher and Eisenman 1988) and leads to an increased life span of these short-lived proteins. Two hnRNP-associated proteins become insoluble after a heat shock of HeLa cells (Lutz et al. 1988). These changes might represent structural modification of these proteins or they might be a consequence of the overall structural changes which appear to increase the protein content of nuclei in heat-shocked cells (Warters et al. 1986; Berrios and Fisher 1988). Another example is the dsRNA-dependent protein kinase induced by interferon in mammalian cells (Dubois et al. 1989). Dubois and coworkers (1987) had previously noticed that some factor leading to the antiviral state induced by interferon treatment was abolished by heat shock. Both the p68 kinase and a 2-5A synthetase are cytoplasmic soluble enzymes involved in the antiviral response and both are thought to be associated with ribosomes. The p68 kinase autophosphorylation but not the 2–5A synthetase activity was missing in lysates from freshly heat-shocked cells. The disappearance of p68 autokinase activity was a rather specific effect since most of the protein phosphorylation which occurred in vitro was not affected by the same heat shock conditions. Cell lysates in nonionic detergent were fractionated by centrifugation and samples were probed by Western blot with p68 kinase specific antibodies. The total amount of p68 in the cell did not vary, indicating no increased degradation. While most of the p68 protein was in the supernatant fraction of control cells it remained in the insoluble pellets from heat-shocked cell lysates, but no relocalization was detected by immunocytochemistry. Meanwhile, the distribution and amounts of 2–5A synthetase protein analyzed by Western blot remained unaffected by stress. The loss of p68 kinase activity was specific and correlated with a loss of solubility of the protein.

Thus after a heat shock various endogenous and two foreign proteins become insoluble, but the extent of this change depends on the protein and may lead to a loss of enzymatic activity.

Heat Shock Proteins Are the Predominant Insoluble Proteins Detected After Stress

What is the extent of protein insolubility in a heat-shocked cell? To address this question, cells were labeled prior to stress, then lysed with a nonionic detergent immediately after stress and fractionated by centrifugation. Only a few differences between stressed and nonstressed cells were detected in the electrophoretic pattern of proteins in the insoluble pellet; the major changes were increased amount of 85– and 70-kDa proteins in the heat-shocked pellet (Fig. 8.1). These proteins comigrated in 2-D polyacrylamide gel electrophoresis with the 90- and

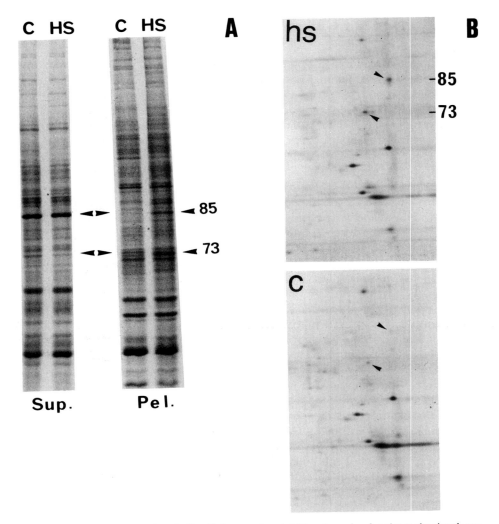

Fig. 8.1. Effect of heat shock on bulk cellular protein solubility. *Arrowheads* point to the signals corresponding to the constitutive heat shock proteins of 73 and 84/86 kDa. Mouse Ltk⁻ fibroblasts grown in DMEM medium supplemented with 10% fetal calf serum were labeled 18 h with ^{35}S-methionine (50 μCi/ml) prior to a 15-min heat shock at 45 °C. Immediately after stress, cells were lysed in Triton buffer (20 mM Tris pH 7.4, 50 mM NaCl, 10 mM $MgCl_2$, 10 mM mercaptoethanol, 1% Nonidet P-40, 10% glycerol). Cell lysates were fractionated into a supernatant (Sup.) and a pellet (Pel.) by centrifugation at 10 000 g for 10 min. **A** One-dimensional gel autoradiography. Supernatants from 2.5×10^4 cells and pellets from 2×10^5 cells were electrophoresed on the same polyacrylamide gel. **B** Two-dimensional gel electrophoresis of the control (*c*) and heat-shocked (*hs*) cell pellets

73-kDa constitutive forms respectively of the heat shock proteins. These latter proteins are among the most abundant proteins in the unstressed cell, where they are mainly distributed in the cytoplasm. The presence of the 90-, and 24-kDa protein families and of ubiquitin in insoluble aggregates has been reported

(Arrigo et al. 1980; Tanguay and Vincent 1982; Collier and Schlesinger 1986; Napolitano et al. 1987; Carlson et al. 1987; Bond et al. 1988), and most studies point to a redistribution of the HSPs after stress or during mild hyperthermia. The 24-kDa chicken heat shock protein aggregates into large cytoplasmic granules (Collier et al. 1988); its corresponding mammalian homolog fractionated with the nucleus and became detergent-insoluble (Arrigo et al. 1988). Proteins of the 70-kDa family migrate into the nucleus and concentrate within the nucleolus (Pouchelet et al. 1983; Velazquez and Lindquist 1984). However, in the experiment described in Fig. 8.1, the duration of stress was short and did not allow for de novo synthesis of HSPs. As already reported by Lewis and Pelham (1985) in a similar protocol, little or no spatial redistribution of the 70-kDa HSP was detected immediately after stress and translocation into the nucleoli took place at longer times, during a recovery period (Mattei et al. 1989).

Protein Insolubilization Corresponds to a Denaturation

Thermal denaturation has been proposed to be the rate-limiting step in hyperthermic cell killing (Rosenberg et al. 1971). This hypothesis has been supported by studies on effects of solvents on hyperthermic cell killing. Addition of monohydric alcohols such as ethanol to the culture medium sensitizes cells to heating and favors thermal denaturation of proteins in vitro, whereas addition of glycerol or replacement of water by deuterium oxide protects cells (Massicotte-Nolan et al. 1981). It was reported recently that protein denaturation occurs in hamster cells at temperatures slightly above 37°C (Lepock et al. 1988). The onset temperatures for protein denaturation are lower in the presence of ethanol but increase in the presence of glycerol or deuterium oxide. Similar data are obtained from studies of enzyme solubility. At 37°C, ethanol decreases luciferase solubility (Nguyen et al. 1989), but conversely glycerol and deuterium oxide protect luciferase, β-galactosidase, p68 kinase, and the 90- and the 70-kDa constitutive heat shock proteins against loss of enzymatic activity and changes in solubility (unpubl. data).

Thermotolerance Is Linked to Decrease in Protein Insolubility

Thermotolerant cells recover their cytoskeleton morphology and normal macromolecule synthesis faster than nontolerant cells (Sciandra and Subjeck 1984; Yost and Lindquist 1986; Welch and Mizzen 1988; Black and Subjeck this vol.). These observations lead to the following question: are thermotolerant cells less damaged or do they have a higher efficiency of repair? The reporter enzymes, p68 kinase, luciferase, and β-galactosidase were found to be less heat-sensitive after a "priming" stress (Nguyen et al. 1989; M F Dubois unpubl. data). This protection was detected and extended with other stress agents: luciferase was less ethanol-sensitive after a priming heat shock. Thus we propose that thermotolerant cells will be more resistant and recover faster because they are less damaged. Indeed, the collapse of the intermediate filament cytoskeleton and the alteration in snRNP structure are attenuated in the thermotolerant cell (Welch and Mizzen 1988). Thus it appears, as a rule, that hyperthermia-induced damages in cells are minimized in a thermotolerant cell.

A "priming" stress also attenuates the inhibition of protein synthesis which occurs during and after heat shock (Sciandra and Subjeck 1984; Mizzen and Welch 1988). We have found that thermoprotective effects on protein synthesis required stronger "priming" stress than the thermoprotective effects on luciferase activity (Nguyen et al. 1989). This suggests that different processes are involved in the protection of these systems. If HSPs are involved, this discrepancy may reflect the noncoordination of their synthesis after stress. Many experiments have suggested that thermotolerance is related to the expression of heat shock proteins (Li and Werb 1982; Subjeck and Sciandra 1982; Landry et al. 1982; Carper et al. 1987). However, data have also been obtained which conflict with this view (Hall 1983; Hallberg 1986; Widelitz et al. 1986; Van Bogelen et al. 1987).

Interpretation of the Behavior of Heat Shock Proteins During Heat Shock

At first glance, it appears paradoxical that members of the heat shock protein family are the major cellular proteins to become insoluble during heat shock. Stabilization of stress-susceptible proteins was expected to be provided by stress-resistant proteins (Minton et al. 1982).

We interpret this observation with a model based on the chaperonin activities of heat shock proteins (Finley et al. 1984; Pelham 1986; Pelham 1988; Bochkareva et al. 1988; Ellis and Hemmingsen 1989). Unfolding of cellular proteins during heat shock should expose hydrophobic internal domains which would bind several constitutive HSP molecules. This binding is reversed by ATP hydrolysis in those complexes containing the 60-and 70-kDa family. Thus, under stress conditions, an increasing number of HSPs would be required to prevent aggregation of the unfolded polypeptides. After some time of stress the pool of "available" HSP molecules would be depleted. The consequence is that some "sticky" domains of unfolded proteins will continue to bind HSPs, while others, unprotected, will start to promote aggregation and drive along the bound HSP molecules. Increasing the pool of "available" HSP molecules would prevent the aggregation of unfolded polypeptides and account for thermoprotection of enzymes, while decreasing the pool of HSP molecules should result in higher thermosensitivity. HSPs of the 70-kDa family start to aggregate in vitro upon heating at stress temperatures (Sadis et al. 1990). This self-aggregation, which is prevented by ATP, may worsen the above-mentioned depletion.

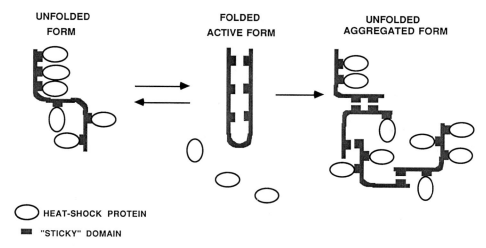

| UNFOLDED FORM | FOLDED ACTIVE FORM | UNFOLDED AGGREGATED FORM |

○ HEAT-SHOCK PROTEIN

■ "STICKY" DOMAIN

Fig. 8.2. Proposed mechanism for insolubilization of heat shock proteins as a result of cellular protein aggregation. Cellular proteins are expected to show a conformational equilibrium between folded and partially unfolded structures. Heat shock proteins are expected to act as chaperonins, binding to the unfolded forms and preventing their aggregation. This binding is reversible (after ATP hydrolysis for HSPs of the 60- and 70-kDa family). Heat shock or addition of solvents such as ethanol favors the unfolded conformations and aggregation occurs at a faster rate than "chaperoning". In this case the most thermosensitive cellular proteins start to aggregate. Increasing the pool of free HSPs by a "priming" stress favors "chaperoning", which delays aggregation

We may also understand why the in vivo denaturing temperature for a given protein depends on the cell in which it is expressed. For instance, firefly lucifer-ase reporter was found to be less thermostable within an insect (*Drosophila*) cell than within a mammalian (mouse) cell (Nguyen et al. 1989). The heat shock response is triggered by lower temperatures in *Drosophila* than in mouse cells; therefore depletion of available heat shock proteins should occur at lower temperatures. Furthermore, increasing the level of heat shock proteins present under normal growth conditions in a given cell should increase the pool of "available" HSP and prevent the aggregation of unfolded species. Such are the results presented by Gatenby and coworkers: many thermosensitive mutations which should correspond to less thermostable proteins can be rescued at nonpermissive temperature in bacteria by overproduction of the GroE HSPs (see Gatenby et al. this vol.). In particular, the folding mutant tsfH304 tailspike from *Salmonella* phage P22 is rescued at nonpermissive temperature. Aggregation of phage P22 protein has been proposed as a model for inclusion body formation. The unfolded native protein is matured as a trimer which is very thermostable. At nonpermissive temperature the formation of trimers is impaired and the remaining native polypeptides aggregate into inclusion bodies (Haase-Pettingell and King 1988). Thus, excess GroE proteins could chaperon more efficiently unfolded polypeptides and prevent inclusion body formation. More generally, overproduction of GroE HSPs under nonstress conditions brings thermoresistance to *E. Coli* cells (Kusukawa and Yura 1988). Synthesis of *Drosophila* HSP70 accelerates the recovery of nucleolar morphology of heat-shocked mammalian cells (Pelham 1984). Overproduction of the 27-kDa HSP brings thermoresistance to mammalian cells (Landry et al. 1989). However, one question remains: if overexpression of individual HSPs under standard conditions brings thermoresistance, why is overexpression of all the HSPs under nonstress conditions inefficient (Van Bogelen et al. 1987)?

Heat Shock Protein Insolubility and Heat Shock Gene Activation

Several observations suggest that the appearance of "abnormal" proteins triggers heat shock protein synthesis (Hightower 1980; Goff and Goldberg 1985; Ananthan et al. 1986; Edington et al. 1989). Solvents which influence protein denaturation in vitro affect similarly the synthesis of heat shock proteins in vivo. Thus, it had been previously proposed that saturation of the *E.Coli* proteolytic degradation pathway or depletion of free ubiquitin in eukaryotic cells might signal heat shock gene induction (Munro and Pelham 1985). However, this hypothesis received little experimental support because free ubiquitin does not decrease significantly during stress (Bond et al. 1988). In *E.Coli* a heat shock response is induced by unfolded proteins and depends on accumulation of unfolded proteins rather than saturation of the degradation pathways (Parsell and Sauer 1989). We propose to extend the ubiquitin depletion hypothesis: depletion of various "available" heat shock proteins might be at the basis of the heat shock gene activation. Conversely, excess "available" heat shock proteins would inhibit the expression of the corresponding genes and account for the observed apparent self-regulation (DiDomenico et al. 1982).

References

Ananthan J, Goldberg AL, Voellmy R (1986) Abnormal proteins serve as eukaryotic stress signals and trigger the activation of heat shock genes. Science 232:522–524

Anderson RL, Hahn GM (1985) Differential effects of hyperthermia on the Na+, K+-ATPase of Chinese hamster ovary cells. Radiat Res 102:314–323

Arrigo AP (1983) Acetylation and methylation patterns of core histones are modified after heat or arsenite treatment of *Drosophila* tissue culture cells. Nucl Acid Res 11:1389–1404

Arrigo AP, Fakan S, Tissières A (1980) Localization of the heat shock-induced proteins in *Drosophila melanogaster* tissue culture cells. Dev Biol 78:86–103

Arrigo AP, Suhan JP, Welch WJ (1988) Dynamic changes in the structure and intracellular locale of mammalian low-molecular-weight heat shock protein. Mol Cell Biol 8:5059–5071

Berrios S, Fisher PA (1988) Thermal stabilization of putative karyoskeletal protein-enriched fractions from *Saccharomyces cerevisiae*. Mol Cell Biol 8:4573–4575

Black AR, Subjeck JR (1989) Involvement of rRNA synthesis in the enhanced survival and recovery of protein synthesis seen in thermotolerance. J Cell Physiol 138:439–449

Bochkareva ES, Lissin NM, Girshovich AS (1988) Transient association of newly synthesized unfolded proteins with the heat-shock GroEL protein. Nature (Lond) 336:254–257

Bond U, Schlesinger MJ (1985) Ubiquitin is a heat shock protein in chicken embryo fibroblasts. Mol Cell Biol 5:949–956

Bond U, Agell N, Haas LH, Redman K, Schlesinger MJ (1988) Ubiquitin in stressed chicken embryo fibroblasts. J Biol Chem 263:2384–2388

Burdon RH, Gill VM, Rice-Evans C (1987) Oxidative stress and heat shock protein induction in human cells. Free Rad Res Comms 3:129–139

Calderwood SK, Hahn GM (1983) Thermal sensitivity and resistance of insulin-receptor binding. Biochim Biophys Acta 756:1–8

Camato R, Tanguay RM (1982) Changes in the methylation pattern of core histones during heat shock in *Drosophila* cells. EMBO J 1:1529–1532

Carlson N, Rogers S, Rechsteiner M (1987) Microinjection of ubiquitin: changes in protein degradation in HeLa cells subjected to heat shock. J Cell Biol 104:547–555

Carper SW, Duffy JJ, Gerner EW (1987) Heat shock proteins in thermotolerance and other cellular processes. Cancer Res 47:5249–5255

Cheng KH, Hui SW, Lepock JR (1987) Protection of the membrane calcium adenosine triphosphatase by cholesterol from thermal inactivation. Cancer Res 47:1255–1262

Chrétien P, Landry J (1988) Enhanced constitutive expression of 27-kDa heat shock proteins in heat-resistant variants from Chinese hamster cells. J Cell Physiol 137:157–166

Collier NC, Schlesinger MJ (1986) The dynamic state of heat shock proteins in chicken embryo fibroblasts. J Cell Biol 103:1495–1507

Collier NC, Heuser J, Aach Levy M, Schlesinger MJ (1988) Ultrastructure and biochemical analysis of the stress granule in chicken embryo fibroblasts. J Cell Biol 106:1131–1139

DiDomenico BJ, Bugaisky GE, Lindquist S (1982) The heat shock response is self-regulated at both the transcriptional and posttranscriptional levels. Cell 31:593–603

Drummond IAS, Steinhardt RA (1987) The role of oxidative stress in the induction of *Drosophila* heat shock proteins. Exp Cell Res 173:439–449

Dubois MF, Ferrieux C, Robert N, Lebon P, Hovanessian AG (1987) Modification, after heat shock, of the antiviral state induced by interferon in murine L cells. Ann Inst Pasteur/Virol 138:345–353

Dubois MF, Galabru J, Lebon P, Safer B, Hovanessian AG (1989) Reduced activity of the interferon-induced double-stranded RNA-dependent protein kinase during a heat shock stress. J Biol Chem 264:12165–12171

Duncan R, Milburn SC, Hershey JWB (1987) Regulated phosphorylation and low abundance of HeLa cell initiation factor eIF-4F suggest a role in translational control. J Biol Chem 262:380–388

Edington BV, Whelan SA, Hightower LE (1989) Inhibition of heat shock (stress) protein induction by deuterium oxide and glycerol: additional support for the abnormal protein hypothesis of induction. J Cell Physiol 139:219–228

Ellis RJ, Hemmingsen SM (1989) Molecular chaperones: proteins essential for the biogenesis of some macromolecular structures. Trends Biochem Sci 14:339–342

Evan GI, Hancock DC (1985) Studies on the interaction of the human c-myc protein with cell nuclei: p62c-myc as a member of a discrete subset of nuclear proteins. Cell 43:253–261

Finley D, Ciechanover A, Varshavsky A (1984) Thermolability of ubiquitin-activating enzyme from the mammalian cell cycle mutant ts85. Cell 37:43–55

Glover CVC, Vavra KJ, Guttman SD, Gorovsky MA (1981) Heat shock and deciliation induce phosphorylation of histone H1 in *T.pyriformis*. Cell 23:73–77

Goff SA, Goldberg AL (1985) Production of abnormal proteins in E. coli stimulates transcription of lon and other heat shock genes. Cell 41:587–595

Haase-Pettingell CA, King J (1988) Formation of aggregates from a thermolabile in vivo folding intermediate in P22 tailspike maturation. J Biol Chem 263:4977–4983

Hall BG (1983) Yeast thermotolerance does not require protein synthesis. J Bacteriol 156:1363–1365

Hallberg RL (1986) No heat shock protein synthesis is required for induced thermostabilization of translational machinery. Mol Cell Biol 6:2267–2270

Hightower LE (1980) Cultured animal cells exposed to amino acid analogues or puromycin rapidly synthesize several polypeptides. J Cell Physiol 102:407–427

Kane JF, Hartley DL (1988) Formation of recombinant protein inclusion bodies in *Escherichia coli*. Trends Biotech 6:95–101

Kennedy IM, Burdon RH, Leader DP (1984) Heat shock causes diverse changes in the phosphorylation of the ribosomal proteins of mammalian cells. FEBS Lett 169:267–273

Kusukawa N, Yura T (1988) Heat shock protein GroE in *Escherichia coli*: key protective roles against thermal stress. Genes Dev 2:874–882

Landry J, Bernier D, Chrétien P, Nicole LM, Tanguay RM (1982) Synthesis and degradation of heat shock proteins during development and decay of thermotolerance. Cancer Res 61:428–437

Landry J, Chrétien P, Lambert H, Hichey E, Weber LA (1989) Heat shock resistance conferred by expression of the human HSP27 gene in rodent cells. J Cell Biol 109:7–15

Lepock JR, Frey HE, Rodahl AM, Kruuv J (1988) Thermal analysis of CHL V79 cells using differential scanning calorimetry: implications for hyperthermic cell killing and the heat shock response. J Cell Physiol 137:14–24

Lewis MJ, Pelham HRB (1985) Involvement of ATP in the nuclear and nucleolar functions of the 70 kd heat shock protein. EMBO J 4:3137–3143

Li GC, Werb A (1982) Correlation between synthesis of heat shock proteins and development of thermotolerance in Chinese hamster fibroblasts. Proc Natl Acad Sci USA 79:3218–3222

Littlewood TD, Hancock DC, Evan GI (1987) Characterization of a heat shock-induced insoluble complex in the nuclei of cells. J Cell Sci 88:65–72

Lutz Y, Jacob M, Fuchs JP (1988) The distribution of two hnRNP-associated proteins defined by a monoclonal antibody is altered in heat-shocked HeLa cells. Exp Cell Res 175:109–124

Lüscher B, Eisenman RN (1988) c-myc and c-myb protein degradation: effect of metabolic inhibitors and heat shock. Mol Cell Biol 8:2504–2512

Magun BE, Fennie CW (1981) Effects of hyperthermia on binding, internalization, and degradation of epidermal growth factor. Radiat Res 86:133–146

Maridonneau-Parini I, Clerc J, Polla BS (1988) Heat shock inhibits NADPH oxidase in human neutrophils. Biochem Biophys Res Commun 154:179–186

Massicotte-Nolan P, Glofcheski DJ, Kruuv J, Lepock JR (1981) Relationship between hyperthermic cell killing and protein denaturation by alcohols. Radiat Res 87:284–299

Mattei D, Scherf A, Bensaude O, Pereira da Silva L (1989) A heat-shock protein from the human malaria parasite *Plasmodium falciparum* induces autoantibodies. Eur J Immunol 19:1823–1828

Minton KW, Karmin P, Hahn GM, Minton AP (1982) Nonspecific stabilization of stress-susceptible proteins by stress-resistant proteins: a model for the biological role of heat shock proteins. Proc Natl Acad Sci USA 79:7107–7111

Mivechi NF, Dewey WC (1985) DNA polymerase α and β activities during the cell cycle and their role in heat radiosensitization in Chinese hamster ovary cells. Radiat Res 103:337–350

Mizzen LA, Welch WJ (1988) Characterization of the thermotolerant cell. I. Effects on protein synthesis activity and the regulation of heat-shock protein 70 expression. J Cell Biol 106:1105–1116

Munro S, Pelham H (1985) What turns on heat shock genes? Nature (Lond) 317:477–478

Napolitano EW, Pachter JS, Liem RKH (1987) Intracellular distribution of mammalian stress proteins. J Biol Chem 262:1493–1504

Nguyen VT, Morange M, Bensaude O (1989) Protein denaturation during heat shock and related stress. J Biol Chem 264:10487–10492

Nolan NL, Kidwell WR (1982) Effect of heat shock on poly(ADP-ribose) synthetase and DNA repair in *Drosophila* cells. Radiat Res 90:187–203

Pain R (1987) Protein folding for pleasure and for profit. Trends Biochem Sci 12:309–312

Parag HA, Raboy B, Kulka RG (1987) Effect of heat shock on protein degradation in mammalian cells: involvement of the ubiquitin system. EMBO J 6:55–61

Parker KA, Bond U (1989) Analysis of pre-rRNAs in heat-shocked HeLa cells allows identification of the upstream termination site of human polymerase I transcription. Mol Cell Biol 9:2500–2512

Parsell DA, Sauer RT (1989) Induction of a heat shock-like response by unfolded protein in *Escherichia coli*: dependence on protein level not protein degradation. Genes Dev 3:1226–1232

Pelham HRB (1984). Hsp70 accelerates the recovery of nucleolar morphology after heat shock. EMBO J 3:3095–3100

Pelham HRB (1986) Speculations on the functions of the major heat shock and glucose-regulated proteins. Cell 46:959–961

Pelham HRB (1988) Heat shock proteins: coming in from the cold. Nature (Lond) 332:776–777

Pouchelet M, St-Pierre E, Bibor-Hardy V, Simard R (1983) Localization of the 70 000 dalton heat-induced protein in the nuclear matrix of BHK cells. Exp Cell Res 149:451–459

Ropp M, Courgeon AM, Calvayrac R, Best-Belpomme M (1983) The possible role of the superoxide ion in the induction of heat-shock and specific proteins in aerobic *Drosophila* cells during return to normoxia after a period of anaerobiosis. Can J Biochem Cell Biol 61:456–461

Rosenberg B, Kemeny G, Switzer RC, Hamilton TC (1971) Quantitative evidence for protein denaturation as the cause of thermal death. Nature (Lond) 232:471–473

Sadis S, Raghavendra K, Schuster TM, Hightower LE (1990) Biochemical and biophysical comparison of bacterial DnaK and mammalian hsc73, two members of an ancient stress protein family. In: Villa Franca JJ (ed) Current research in protein chemistry, Academic Press, Lond NY

Sciandra JJ, Subjeck JR (1984) Heat shock proteins and protection of proliferation and translation in mammalian cells. Cancer Res 44:5188–5194

Spiro JJ, Denman DL, Dewey WC (1982) Effect of hyperthermia on CHO DNA polymerases α and β. Radiat Res 89:134–149

Subjeck JR, Sciandra JJ (1982) Coexpression of thermotolerance and heat-shock proteins in mammalian cells. In: Heat Shock From Bacteria to Man. Schlesinger MJ, Ashburner M, Tissières A (eds.) Cold Spring Harbor Lab, Cold Spring Harbor, NY, pp 405–411

Tanguay RM, Vincent M (1982) Intracellular translocation of cellular and heat shock induced proteins upon heat shock in *Drosophila* Kc cells. Can J Biochem 60:306–315

VanBogelen RA, Acton MA, Neidhardt FC (1987) Induction of the heat shock regulon does not produce thermotolerance in *Escherichia coli*. Genes Dev 1:525–531

Velazquez JM, Lindquist S (1984) Hsp 70: nuclear concentration during environmental stress and cytoplasmic storage during recovery. Cell 36:655–662

Warters RL, Brizgys LM, Sharma R, Roti Roti JL (1986) Heat shock (45 °C) results in an increase of nuclear matrix protein mass in HeLa cells. Int J Radiat Biol 50:253–268

Welch WJ, Mizzen LA (1988) Characterization of the thermotolerant cell. II. Effects on the intracellular distribution of heat-shock protein 70, intermediate filaments, and small nuclear ribonucleoprotein complexes. J Cell Biol 106:1117–1130

Welch WJ, Suhan JP (1986) Cellular and biochemical events in mammalian cells during and after recovery from physiological stress. J Cell Biol 103:2035–2052

Westwood JT, Steinhardt RA (1989) Effects of heat and other inducers of the stress response on protein degradation in Chinese hamster and *Drosophila* cells. J Cell Physiol 139:196–209

Widelitz RB, Magun BE, Gerner EW (1986) Effects of cycloheximide on thermotolerance expression, heat shock protein synthesis, and heat shock protein mRNA accumulation in rat fibroblats. Mol Cell Biol 6:1088–1094

Wolffe AP, Perlman AJ, Tata JR (1984) Transient paralysis by heat shock of hormonal regulation of gene expression. EMBO J 3:2763–2770

Yost HJ, Lindquist S (1986) RNA splicing is interrupted by heat shock and is rescued by heat shock protein synthesis. Cell 45:185–193

CHAPTER 9

Mechanisms of Stress-Induced Thermo- and Chemotolerances

Adrian R. Black and John R. Subjeck

Division of Radiation Biology, Roswell Park Cancer Institute, Buffalo, New York 14263, USA

Thermotolerance and Enhanced Recoveries of Protein and RNA Synthesis

Thermotolerance, the acquisition of a transiently increased resistance to heat induced by exposure to elevated temperature was first observed in mammalian cells subjected to fractionated hyperthermia in the treatment of cancer (Gerner and Schneider 1975; Henle and Leeper 1976). The increased resistance is marked; a preconditioning heat treatment can result in increased survival following severe heat treatment in the order of 10^4 to 10^5. This phenomenon is distinct from a clonal selection of resistant cells, as often seen with drug resistance, since it is noninheritable and can be induced by non-lethal preconditioning treatments. Thermotolerance appears to be a universal phenomenon, being observed in all organisms tested (reviewed in Lindquist and Craig 1988; Subjeck and Shyy 1986). The expression of thermotolerance is also seen at the level of the whole organism and occurs under natural conditions as well as following laboratory manipulations (e.g., Easton et al. 1987). Despite its universal nature, little is known of the mechanisms which lead to the enhanced survival of thermotolerant cells.

Another universally observed response of organisms to heat is the induction of a specific set of proteins (reviewed in Lindquist and Craig 1988). These heat shock proteins (HSP) fall into five families with molecular weights of 15–30 kDa (low molecular weight HSP), ~60 kDa (HSP60), ~70 kDa (HSP70), ~90 kDa (HSP90) and 100–110 kDa (HSP110) (Lindquist and Craig 1988). As might be expected from their common induction in all organisms following heat treatment, these proteins are conserved throughout evolution (see Burdon 1986; Lindquist and Craig 1988). In fact, many represent the most highly conserved proteins examined to date: e.g., there is >50% homology between HSP70 in species as divergent as *Escherichia coli* and *Homo sapiens* (Hunt and Morimoto 1985). Generally, heat shock proteins are present as multigene families, with members expressed in unstressed cells as well as following heat treatment (see Lindquist and Craig 1988). For the purpose of this report, the individual

Stress Proteins
Schlesinger, Santoro, Garaci (Eds.)
© Springer-Verlag Berlin Heidelberg 1990

members of particular groups of heat shock proteins will not generally be distinguished, but will be referred to collectively (i.e., as HSP70, HSP90 etc.). There is another closely related set of stress proteins that are induced by stresses such as glucose deprivation and anoxia. These "glucose-regulated proteins" will be discussed in the second part of this chapter. The universal appearance of heat shock proteins, together with their remarkable evolutionary conservation, has led to the hypothesis that they play significant roles in the normal metabolism of the cell, as well as following heat shock.

Although no roles for the low molecular weight HSP and HSP110 have been identified, recently much information has been gathered regarding the normal cellular functions of HSP70 and HSP90. The HSP70s are ATP-binding proteins (Welch and Feramisco 1985) and have ATPase activity. In mammalian cells, an HSP70 has been identified as the clathrin uncoating ATPase (Ungewickel 1985; Chappell et al. 1986). Another role for HSP70 is in transmembrane trafficking of proteins (Deshaies et al. 1988; Chirico et al. 1988). It has been postulated that proteins have to be in an unfolded state to cross cellular membranes and that HSP70 acts as an ATP-dependent "unfoldase", thus enabling their passage across the lipid bilayer.

Heat shock proteins also bind to various factors involved in controlling the growth state of the cell. HSP90 is complexed with the steroid receptor, maintaining the receptor in the inactive state pending agonist binding (Joab et al. 1984; Sanchez et al. 1987). An association of HSP90, which is itself phosphorylated, with many protein kinases has been observed. These include the tyrosine kinase products of the oncogenes src, yes, fps, fes and fgr (Ziemiecki et al. 1986; Brugge et al. 1981; Oppermann et al. 1981) as well as casein kinase II and eIF2α kinase (Rose et al. 1987). HSP70 is associated with and stabilizes p53, (Hinds et al. 1987). Further, information on the proposed roles of heat shock proteins and on genetic analysis of their function can be found elsewhere in this Volume and in reviews by many authors (e.g., Lindquist and Craig 1988; Tomosovic 1989).

The prompt appearance of heat shock proteins following heat shock led to the proposition that they are involved in a homeostatic response, helping cells to recover from the damaging effects of heat and protecting the cells against further insult. A large body of circumstantial evidence has been amassed to support this idea. Several heat-resistant cell lines have been found to have elevated levels of heat shock proteins (Laszlo and Li 1985; Yaraha et al. 1986; Chétien and Landry 1988). Furthermore, it has been demonstrated that when HSP70 synthesis is blocked by transfection with DNA-containing sequences that compete for transcription factors or following microinjection of specific anti-HSP70 antibodies, cells become more heat-sensitive (Riabowol et al. 1988; Johnston and Kucey 1988). Another indication of the role of heat shock proteins in heat protection is the correlation of their levels with thermotolerance. Many investigators have found that the elevated synthesis of heat shock proteins correlates with thermotolerance in prokaryotes and eukaryotes (see Subjeck and Shyy 1986). Closer examination in mammalian cells has found that the level of HSP70 correlates with the level of thermotolerance during both its induction and decay (Subjeck et al. 1982; Li 1985). This relationship is independent of the original stress, since treatments other than heat which induce heat shock proteins, e.g., sodium arsenite and ethanol, also induce thermotolerance (Li 1983). Further, induction of aberrant heat shock proteins by treatment with amino acid analogs does not lead to thermotolerance and even blocks its expression (Li and Laszlo 1985), suggesting that functional HSPs are required. It should be noted, however, that amino acid analogs affect all newly synthesized cellular proteins and may sensitize them to the effects of heat, thus masking any effects of thermotolerance.

Despite the identification of functions for heat shock proteins in normal cellular activities, little is known of the roles of heat shock proteins in heated cells or of the mechanisms underlying acquired thermotolerance. However, by extrapolation of the functions of heat shock proteins in normal cells and the effects of heat shock on cells, theories can been postulated. For example, the ATPase activity of HSP70 together with the discovery that the closely related glucose-regulated protein of 78 kDa is the Bip protein of lymphocytes (this protein binds to heavy chains in immature lymphocytes, preventing their aggregation before they can be linked to light chain; see below), has led to the proposal that these proteins function to either prevent damaging aggregation of proteins following heat shock or help in removing such aggregates once they have formed (Munro and Pelham 1986; Lewis and Pelham 1985). Further, the observed location of HSP110 in the nucleolus (Subjeck et al. 1983), the relocation of HSP70 into the nucleus and in particular the nucleolus (Welch and Suhan 1985) and the apparent relocation of low molecular weight HSPs into the nucleus following heat shock (Arrigo et al. 1988) argue that all are involved in repairing/limiting heat damage in these organelles. The idea that HSP70 is involved in the recovery of nucleolar function (ribosomal RNA synthesis and processing) is strengthened by the observation that, in cells which express elevated levels of this protein, nucleolar morphology recovers more rapidly following heat treatment than in control cells (Pelham 1984). Also, the apparent association of HSP70 and HSP90 with components involved in protein synthesis, argues that these proteins are involved with translation following heat shock (Welch and Suhan 1985; Rose et al. 1989a).

A major problem in determining the molecular basis underlying thermotolerance is the multiple effects of heat shock: heat disrupts the plasma membrane and membrane protein functions, the cytoskeleton, protein synthesis, DNA synthesis, and RNA synthesis and processing (reviewed in Roti-Roti and Laszlo 1988). The pleiotropic nature of a heat shock also leads to the possibility that the importance of various structures/functions to cell killing and thermotolerance may vary depending on the precise conditions. Indeed, this would account for many of the apparent contradictions present in the literature.

One approach to this problem is to (1) study the effects of heat shock on cellular processes, (2) determine how these effects are changed by thermotolerance, (3) determine how these changes correlate with enhanced cell survival, and (4) use agents which specifically disrupt processes affected by heat and thermotolerance in a manner that can reverse or enhance the effects of thermotolerance and thus determine the causal relationship between these observed changes and enhanced cell survival.

Two processes which are dramatically affected by heat are protein and RNA synthesis. The involvement of these processes in heat shock and thermotolerance will be considered mainly in mammalian cells since this is the area of interest of the authors.

Protein Synthesis

Heat shock can cause a severe inhibition in normal protein synthesis; however, the particular effect depends on the temperature used. In mammalian cells, more extreme heat shock temperatures cause a general inactivation of the translational machinery. It has been shown in L5178Y murine leukemia cells that the extent of inhibition of protein synthesis increases with increasing temperature up to 44 °C (Fuhr 1974). Lower temperature heat shocks (41–42 °C) lead to a transient inhibition of protein synthesis which recovers immediately on return to 37 °C and partially, with time, even at the elevated temperature (Goldstein and Penman 1973). Thermotolerance can develop during continuous exposure to 42 °C (Tomasovic 1989), but higher temperatures (>42.5 °C) that do not allow the development of thermotolerance at the elevated temperature (Tomasovic 1989) cause an almost complete inhibition of protein synthesis. After these higher temperature treatments, some time of incubation at 37 °C is required before protein synthesis will recover. Henle and Leeper (1979) demonstrated that the length of this recovery period is dependent on the severity of the heat treatment. They studied the effects of a 45 °C heat shock on ^{35}S-methionine incorporation by Chinese hamster ovary (CHO) cells. Both a 10-min and a 17.5-min heat treatment led to greater than 95% inhibition of protein synthesis; however, it required 4 h at 37 °C to recover from the 10-min treatment, whereas 26 h was required following 17.5 min at 45 °C. This finding has subsequently been confirmed in our laboratory (Black and Subjeck 1986, 1989; Table 9.1) and related to cell survival. There was an inverse correlation between cell survival and the time required for protein synthesis recovery following various exposures of CHO cells to 45 °C. However, the extent of recovery of protein synthesis itself was not an indication of cell survival, as measured in a colony survival assay, since synthesis recovered to control levels even when the majority of cells were incapable of forming colonies (Table 9.1). Rather, it was the time required for recovery of synthesis that was related to survival. This result emphasizes the difference between the short-term integrity of cells and their ultimate viability.

Several groups, including ours, have reported essentially identical results on the effects of thermotolerance on the recovery of protein synthesis (Petersen and Mitchel 1981; Subjeck and Shyy 1986; Black and Subjeck 1986, 1989; Mizzen and Welch 1988). Although the different conditions and cell lines used led to different delays in protein synthesis, all found that thermotolerant cells show an enhanced recovery of protein synthesis following heat shock. Interestingly, this enhanced recovery of protein synthesis following heat treatment could be in-

Table 9.1. Correlation of the rates of recovery of protein and RNA synthesis with thermotolerance

Treatment	Delay in protein synthesis (h)	Delay in RNA synthesis (h)	Survival (%)
a) 5 min, 45 °C	2–4	<1	90–102
12 min, 45 °C	6	6–8	50–60
27 min, 45 °C	14–16	16–18	1–3
b) 1 h interval	10–12	12–14	10–20
4 h interval	6–8	8	50–70
24 h interval	6–8	6–10	50–70

Protein sysnthesis, RNA synthesis and survival were monitored by ^{35}S-methionine incorporation, ^{3}H-uridine incorporation and colony formation, respectively, after either a single 45 °C heat treatment (a) or a 27-min 45 °C heat treatment administered at various times after a preconditioning 5-min, 45 °C treatment. The delay in synthesis signifies the time required for incorporation to reach≥90% of control levels.

Note the same correlation between survival and the recoveries of synthesis in both thermotolerant cells in (b) (those given the preconditioning nonlethal treatment) and the nontolerant cells in (a). Also, although recovery of RNA synthesis required slightly longer than that of protein synthesis, it was more gradual, with early stages in its recovery often preceding those for protein synthesis. Therefore, recovery of RNA synthesis could be a factor in the recovery of protein synthesis, as noted in the text.

duced by other stresses that induced thermotolerance and heat shock proteins, but not by amino acid analogs which induce (presumably nonfunctional) heat shock proteins but not thermotolerance (Mizzen and Welch 1988).

Studies of the relationship between enhanced recovery of protein synthesis and survival of thermotolerant cells (Black and Subjeck 1986, 1989, Table 9.1) showed an inverse correlation between the delay in protein synthesis and the survival of cells during the development of thermotolerance: i.e., as thermotolerance developed after the preconditioning treatment there was a concomitant increase in the survival of cells following a challenge treatment and a decrease in the delay of protein synthesis caused by that treatment (Table 9.1). Interestingly, the correlation between the delay in protein synthesis and survival following a 45 °C challenge was the same for both thermotolerant cells and nontolerant cells: treatments which led to the same survival in each led to the same delay in protein synthesis in each case. For example, both a 12-min challenge to nontolerant cells and a 27-min challenge to thermotolerant cells led to about a 50% survival and each produced the same 6-8-h delay in protein synthesis. However, a 27-min challenge given to cells 1 h after the preconditioning treatment, when thermotolerance was only partially developed, led to a survival and a delay in protein synthesis intermediate between this level and that seen in nontolerant cells following the same challenge (Table 9.1). This clear correlation between survival and the period of delay in protein synthesis in normal and thermotolerant cells argues that reduced delay in protein synthesis is intimately involved with thermotolerance.

In addition to the clearly demonstrated correlation between survival, thermotolerance, and delay in protein synthesis, a preconditioning treatment (heat treatment or other treatment which caused the induction of HSP and thermotolerance) also caused a significant reduction in the inhibition in protein synthesis observed immediately after the challenge treatment (Laszlo 1989). However, the protection correlated with the period of induction of heat shock proteins following the first heat shock and not with thermotolerance. A similar, but much smaller effect, has also been observed in our laboratory (Sciandra and Subjeck 1984). Since this immediate protection of protein synthesis was not seen in the studies on the delay in synthesis, it helps to emphasize that the exact conditions and cell lines used in experiments may have a great effect on the results obtained.

Although the above data argue that enhanced recovery of protein synthesis may well be important in thermotolerance, they do not prove a causative relationship. Such a relationship could be established by the use of inhibitors of protein synthesis. We have attempted to accomplish this using cycloheximide. However, a complex interaction of this drug with heat treatments, noted also by others (Henle and Leeper 1982; Sherwood et al. 1987; Lee et al. 1987), precluded meaningful results.

To understand the mechanisms involved in enhanced recovery of protein synthesis, a detailed knowledge of the effects of heat shock on the translational machinery and how these effects are influenced by thermotolerance will be needed. Inhibition of protein synthesis is apparently due to a block in initiation, which could be accounted for by posttranslational modification of a number of the initiation factors, including phosphorylation of eIF2-α and dephosphorylation of eIF-4B and eIF-4E. However, other evidence suggests that these changes may not be the primary cause of the inhibition (Duncan and Hershey 1989).

Use of selective inhibitors of RNA synthesis have implicated the inactivation of an RNA factor in the inhibition of protein synthesis at higher temperatures (Goldstein and Penman 1973). During a 42 °C heat shock to HeLa cells, this factor seems to be involved with the adaptation of the synthetic machinery to elevated temperature rather than the recovery following return to 37 °C. Since there is no delay in synthesis following a 42 °C heat treatment, it is not known whether the recovery of protein synthesis seen following more severe treatments is more closely related to the adaptation of the synthetic machinery which did require RNA synthesis or its resumption after milder stress.

From the data cited earlier, it is obvious that the enhanced recovery of protein synthesis following severe heat shock is intimately involved with thermotolerance. However, the mechanisms underlying this phenomenon and the possible role of heat shock proteins are unknown. Further, a causative relationship has not been established and this relationship may not apply under all conditions (Laszlo 1989).

It should be noted that protein synthesis is the end point of a complex process. A minor disruption of an essential component could have the same apparent effect as a widespread, severe disruption to many components: i.e., both could lead to an inhibition of protein synthesis. Therefore, from the present studies, it is not possible to say whether the damage caused by heat shock to the translational machinery is less in thermotolerant cells or if they experience the same level of damage and are able to repair it more rapidly. Both, or a combination of these possibilities could lead to a more rapid recovery of protein synthesis in thermotolerant cells following heat shock. An increased understanding of the effects of heat shock on the synthetic machinery and the controls exerted under normal and heat shock conditions should help resolve these matters.

RNA Synthesis and Processing

Heat shock inhibits both RNA processing and synthesis, although the former seems to be more susceptible. At temperatures of 42–43 °C synthesis of ribosomal RNA (rRNA) continued at about one third of control levels and messenger RNA (mRNA), transfer RNA (tRNA), and 5S RNA synthesis were not significantly affected (Sadis et al. 1988). On the other hand, processing of rRNA and mRNA was dramatically reduced (Sadis et al. 1988). Although inhibition of processing of RNA is a major effect of heat shock, information is lacking as to the effect of thermotolerance on this phenomenon.

More severe heat treatments than those described above significantly reduce the synthesis of RNA. However, with the treatments normally applied, the incorporation of radiolabeled precursors into RNA is not totally inhibited. For example, 45 °C challenges to CHO cells leads to only an ~70% inhibition of ^{3}H-uridine incorporation into RNA (Henle and Leeper 1979; Black and Subjeck 1989). The time for recovery to control levels of incorporation was dependent on the severity of the heat treatment: longer heat treatments led to an increasingly long recovery period (Henle and Leeper 1979; Black and Subjeck 1989). As with protein synthesis, there was an identical correlation between the time required for recovery of RNA synthesis and cell survival following heat shock to either thermotolerant or nontolerant cells (Black and Subjeck 1989; Table 9.1): i.e., the time required for the recovery of RNA synthesis was reduced in thermotolerant cells in a manner that correlated with their enhanced survival.

RNA is synthesized in eukaryotic cells by three separate RNA polymerases, each responsible for the production of different species of RNA. These are designated RNA polymerase I, responsible for rRNA production, RNA polymerase II, responsible for mRNA synthesis, and RNA polymerase III, which synthesizes tRNA and the 5S rRNA. Each polymerase has its own set of transcription factors and is under a different set of controls. The effect of heat shock on these various polymerases has been studied (Caizergues-Ferrer et al. 1980). RNA polymerase I was most sensitive to heat, whereas polymerase II and III were relatively insensitive and a 42 °C heat treatment affected only rRNA synthesis (Sadis et al. 1988). By the use of low concentrations of actinomycin D ($\leq 0.1 \mu g/ml$), which selectively inhibits rRNA synthesis whilst having little effect on the synthesis of other species both before and after heat treatment (Perry and Kelly 1970; Black and Subjeck 1989), we determined that the inhibition of ^{3}H-uridine incorporation produced by a 45 °C heat shock to CHO cells was due mainly to inhibition of rRNA synthesis. Therefore, the correlation

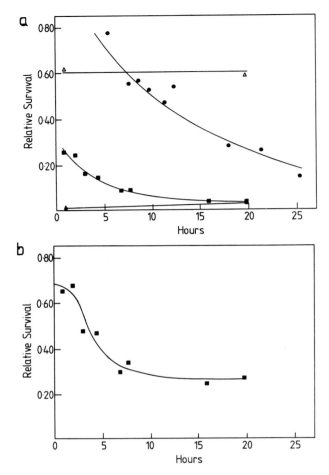

Fig 9.1. The effect of 0.1 μg/ml actinomycin D on the survival of cells following various heat treatments. **a** CHO cells were treated as follows: (△) cells were given a preconditioning 5-min 45 °C treatment, followed by 4 h at 37 °C before a 27-min 45 °C challenge treatment. After the indicated time, cell survival was determined by a colony formation assay. (■) Cells were treated as in (△) except that 0.1 μg/ml actinomycin D was added 1 h prior to the preconditioning treatment and removed when cells were plated for survival assay; (●) actinomycin D (0.1 μg/ml) was added to cells and, after the indicated time at 37 °C, was removed immediately prior to plating for cell survival assay; (▲) cells were given a 27-min 45 °C heat treatment and then incubated for the indicated times at 37 °C before being plated for cell survival assay. **b** Data for cells given a combined heat and drug treatment (■) expressed relative to that for cells given the heat treatment alone (△) and cells given the same drug exposure but no heat treatment (●): i.e., survival for a particular time from [(■) in **a**] was divided by the survival of at that time from [(△) in **a**] and by the survival at that time plus 5.5 h (to compensate for the time actinomycin D was present before 0 time at the end of the challenge treatment) from [(●) in **a**].

Note that although this concentration of actinomycin D was toxic to CHO cells (●), this effect was much less than the toxicity of the challenge treatment to non-tolerant cells (▲) and to the effects of thermotolerance [compare (△) with (▲)]. Following compensation for the toxicity of the heat treatments and the drug treatment (**b**), actinomycin D could be clearly seen to have a time-dependent effect on the expression of thermotolerance. The lower than expected survival seen when the drug was removed immediately after the challenge treatment (0 h in **b**)) can be attributed to the slight synergistic interaction of severe heat treatment with actinomycin D, as noted in the text. Separate experiments determined that addition of the drug after the development of thermotolerance (here it was added prior to its development) made little difference to the inhibition of the expression of thermotolerance (see Black and Subjeck 1989)

of survival following heat treatment and the recovery of RNA synthesis can be extended to a correlation between survival and recovery of rRNA synthesis. This is corroborated by a report by Burdon which states that there is an enhanced recovery of rRNA synthesis in thermotolerant cells (Burdon 1986).

Again, as with the inhibition/recovery of protein synthesis, it is impossible to determine whether the more rapid recovery of RNA synthesis in thermotolerance is due to less damage caused by the heat treatment or whether thermotolerant cells repair more rapidly an equivalent amount of damage (see above).

Protection of rRNA synthesis and processing has been proposed to be one of the functions of heat shock proteins (Pelham 1984). In order to investigate the causative relationship between enhanced recovery of rRNA synthesis and the enhanced survival of thermotolerant cells following heat shock, we undertook a series of studies which used low concentrations of actinomycin D to selectively block the recovery of rRNA synthesis following heat shock (Black and Subjeck 1989). The actions of actinomycin D on RNA synthesis were readily reversible upon removal of the drug (3–5 h). Therefore, by maintaining cells in the presence of the drug for various periods of time after a heat shock, the delay in rRNA synthesis could be manipulated. We found that the drug neither blocked the acquisition of thermotolerance nor the induction of heat shock proteins following a mild heat treatment. Actinomycin D (0.1 μg/ml) was somewhat toxic to CHO cells and did have a slightly synergistic effect in reducing cell survival when applied in combination with a severe heat treatment. However, neither of these effects was influenced by thermotolerance per se and they were small when compared to the enhanced survival produced by thermotolerance. When the effect of delaying the recovery of rRNA synthesis following challenge to thermotolerant cells was examined, actinomycin D was found to have a greater effect than would be expected from its toxicity or synergism with the challenge treatment (Fig. 9.1). This effect had a time course which would be expected if it were due to a blockade of the recovery of RNA synthesis seen in Table 9.1. If left in place for less than 3 h after the challenge treatment, when it would not be expected to have any influence on RNA synthesis (since this would not have recovered), the drug had no effect on relative survival. However, with longer exposure following the challenge treatment, where the drug would delay the recovery of RNA synthesis (note that, since the reversal of the effects of the drug required 3–5 h and the recovery of RNA synthesis required ~8 h, exposures to the drug of greater than 3–5 h would affect the recovery of RNA synthesis), the drug had a progressively greater effect on the expected survival (Fig. 9.1). This effect was seen with drug exposures up to 10 h and reduced the survival to ~40% of that expected from the survival of control cells and the toxicity of the drug. Since the effect could not be accounted for by either the toxicity of the drug or its combined effect with heat shock alone, it was concluded that the drug was able to block the expression of thermotolerance in otherwise thermotolerant cells. The concentration dependence of the effect of actinomycin D on the expression of thermotolerance indicated that it was acting through the selective inhibition of rRNA synthesis; the half maximal dose for inhibition of the expression of thermotolerance (0.01–0.03 μg/ml) was essentially the same as that seen for inhibition of rRNA synthesis (Perry and Kelly 1970). Together with the observation that the vast majority of the inhibition and recovery of RNA synthesis seen in these studies could be accounted for by rRNA synthesis, these data lead to the conclusion that it is the enhanced recovery of rRNA synthesis that is important to the enhanced survival seen in thermotolerance. Therefore, the enhanced recovery of rRNA plays a significant role in the thermotolerance of these cells following the challenge heat treatment. It should be noted that actinomycin D was not able to totally block the expression of thermotolerance in these cells; therefore, it would appear that other factors also play a role in the enhanced survival of the cells. The exact conditions of heat treatment and cell type may alter the degree of relevance of these various parameters to thermotolerance.

The putative RNA factor involved in adaptation of protein synthesis was also sensitive to relatively low concentrations of actinomycin D (Goldstein and Penman 1973), therefore inhibition of the synthesis of this RNA could account for the results seen here. This seems unlikely, since although low concentrations of actinomycin D did cause a delay in the recovery of protein synthesis following challenge to thermotolerant cells (Black and Subjeck 1989), (1) the concentration of actinomycin D needed to inhibit the expression of thermotolerance ($0.01-0.03\,\mu g/ml$, Black and Subjeck 1989) was less than that required to prevent adaptation of synthesis ($0.05-0.1\,\mu g/ml$, Goldstein and Penman 1973), (2) actinomycin D did not affect the recovery of protein synthesis following a 5-min $45\,°C$ heat treatment and (3) protein synthesis did recover to the levels seen in cells treated with actinomycin D alone (Black and Subjeck 1989). Therefore, the most reasonable explanation of these results is that the delay in rRNA synthesis following heat treatment can affect the survival of cells and that this delay is reduced in thermotolerance.

The precise lesions which occur in the RNA polymerase I transcription system as a result of heat shock, the effects of thermotolerance, and possible roles of heat shock proteins are being investigated in this laboratory using an in vitro transcription system.

As mentioned above, actinomycin D treatment delayed the recovery of protein synthesis following challenge to thermotolerant cells. In the presence of the drug, the recovery became more gradual and full recovery was delayed from ~6h post challenge to ~20h (Black and Subjeck 1989). This raises the interesting possibility that it is the delay in protein synthesis that is actually lethal to these cells. If this were the case, enhanced recovery of RNA synthesis would be just one mechanism by which thermotolerant cells are able to reduce the delay in protein synthesis and thus increase their survival after heat shock.

Although increasing amounts of information are becoming available concerning the role of heat shock proteins and the cellular effects of heat shock, little is known about the mechanisms underlying thermotolerance and the role of heat shock proteins in this phenomenon. A large part of this uncertainty can be attributed to the pleiotropic nature of the effects of heat shock and the probability that the exact experimental conditions affect the importance of the various lesions to the ultimate survival of cells. Despite this, studies in our laboratory and others have highlighted the probable role of protection/enhanced recovery of the translational machinery in thermotolerance. Also, we have demonstrated, through the use of a selective inhibitor, that enhanced recovery of RNA, presumably rRNA synthesis, is one mechanism which can lead to enhanced survival of thermotolerant cells.

Glucose-Regulated Proteins and Drug Resistance

It is well known that glucose starvation induces a set of stress proteins referred to as "glucose-regulated" proteins (GRPs, Pouyssegur et al. 1977). A number of other inducers have been identified which also increase the synthetic rates of these stress proteins, including anoxia, calcium ionophore A23187, glucosamine, tunicamycin, and 2-deoxyglucose (Shen et al. 1989; Watowich and Morimoto 1988). Other terminology was subsequently invoked to distinguish this class of proteins (i.e., calcium-regulated proteins, oxygen-regulated proteins); however, to simplify nomenclature we will refer to this set of proteins as initially designated. Three principal glucose-regulated protein species have been widely reported at approximately 78, 97, and 170 kDa. These GRP proteins were initially included in the pool of "stress-inducible" proteins based on their co-induction with the heat shock proteins by certain amino acid analogs. However, the GRPs were unique in that they did not exhibit a strongly increased expression during heat shock; indeed GRPs do not appear to be inducible at all by standard heat shock protocols in higher organisms (the 78-kDa GRP is heat-inducible in yeast, Rose et al. 1989b). The cloning of the principal heat shock- and glucose-regulated genes was a major advance in understanding the relationship between these stress systems. The 78-kDa GRP was identified as a member of the HSP70 family of proteins and the 97-kDa GRP a member of the HSP90 family of proteins (Munro and Pelham 1986; Mazzarella and Green 1987; Sorger and Pelham 1987). The genes for the 110-kDa HSP and the 170-kDa GRP have not yet been cloned.

While there have been an extensive number of reports concerning many aspects of the heat shock system in the last few years, comparatively few studies have appeared which have been directed at the glucose-regulated system of proteins. While we are not reviewing this area here, it is useful to note that GRPs appear to be localized in the endoplasmic reticulum/nuclear envelope compartment (Munro and Pelham 1986: Subjeck and Shyy 1986). The 78-kDa GRP has been identified as the immunoglobulin binding protein in hybridoma cells (Bole et al. 1986; Munro and Pelham 1986). In the endoplasmic reticulum this protein binds the immunoglobulin heavy chain until the light chain is synthesized, at which time the immunoglobulin is assembled and secretion follows. A yeast counterpart to GRP78 has been identified and functions in karyogamy (nuclear fusion) of haploid yeast nuclei, suggesting a function and/or locale in the nuclear envelope. The 97-kDa GRP (also referred to as GRP94, HSP100, or HSP108: this protein is distinct from HSP110) is considered to be a trans-membrane glycoprotein of the endoplasmic reticulum with an exposed cyto-plasmic region (Mazzarella and Green 1987).

In general, a large majority of inducers of the glucose-regulated system affect the glycosylation of proteins in the endoplasmic reticulum (Watowich and Morimoto 1988). Other inducers (e.g., glucose starvation, anoxia, glucosamine) may, in as yet undefined ways, alter protein folding and structure in the endoplasmic reticulum/nuclear envelope compartment, but do not initially alter apparent glycosylation of proteins as observed by endoglycosidase H digestion. Curiously, it has long been recognized that the 97-kDa GRP is a major cellular glycoprotein (Subjeck and Shyy 1986; Mazzarella and Green 1987) and studies with GRP inducers that inhibit glycosylation (e.g., 2-deoxyglucose and tunicamycin) and studies with endoglycosidase H indicate that 170-kDa GRP is also a glycoprotein. Both GRP97 and GRP170 can be induced in either the glycosylated or nonglycosylated form, depending on the inducer. However, whether the glycosylation of the induced form of the GRP is affected or not, a complex between 170-kDa GRP and the 78-kDa GRP is observed (i.e., antibodies prepared against the 170-kDa protein coprecipitate the 78-kDa form, but do not react with this form on Western blots). Indeed, co-precipitation occurs between the 97-kDa GRP and the 78-kDa GRP. The formation of these complexes occurs under GRP-inducing stress conditions only, and GRP78 is not coprecipitated from unstressed, control cells (Zhang et al. 1990). The stress-induced glycosylated form of GRP170 may be altered and in analogy with the proposed function of GRP78 as the immunoglobulin-binding protein, the 170- and 97-kDa proteins are also bound to GRP78 under stress, perhaps to block hydrophobic interactions between nonglycosylated or stress-induced "altered" protein regions. These GRP protein complexes may exhibit some function (as suggested in the case of the GRP78-heavy chain complex) and/or reduce the "active" concentration of GRP78, 97, and 170 in the endoplasmic reticulum/ nuclear envelope compartment during GRP inducing stress.

Since the induction of the glucose-regulated protein system is sensitive to some very interesting physiological conditions, we have focused much of our attention in this area of stress protein biology. Indeed, inducers such as chronic anoxia and glucose starvation are conditions expected to occur during tumor development and increased quantities of GRP78 appear in central regions of some larger tumors (Shen et al. 1987). Resistance phenomena associated with these inducers could be of major importance in determining tumor response to various treatments. Because of the genetic similarities of the major GRPs and HSPs (described above), it would appear reasonable that thermotolerance may occur as a result of GRP induction. Studies of this problem have concluded otherwise: different GRP inducers lead to virtually no thermotolerance and one study demonstrated a modest thermosensitization (Sciandra and Subjeck 1983; Stevenson et al. 1987). The differential localization of the two sets of proteins may explain this if one assumes that the endoplasmic reticulum is not a critical target leading to cell death after a severe heat challenge: the GRPs are confined to the endoplasmic reticulum/nuclear envelope compartment whereas various heat shock proteins are in other cell organelles. In addition, due to their observed interaction, GRP-inducing stress may actually reduce the concentration of "active" GRP monomers in this organelle. Too little is understood concerning the similarities and differences in the functions of these sets of proteins at this time to speculate on the apparent inability of GRPs to confer thermotolerance.

While the induction of GRPs is not associated with thermotolerance, early studies had shown that both anoxia and 2-deoxyglucose lead to substantial resistance to the anti-neoplastic drug adriamycin (Colofiore et al. 1982; Born and Eichholtz-Wirth 1981). We have studied this phenomenon further demonstrating that adriamycin resistance is induced not only by anoxia and 2-deoxyglucose but also by several other GRP inducers (glucosamine, EGTA, glucose starvation, A23187, etc.) and requires conditions/concentrations required for GRP induction (Shen et al. 1987, unpubl. data). Hyperthermia and ethanol also lead to adriamycin resistance (Hahn and Strande 1976; Li and Hahn 1978). The mechanism responsible for drug resistance during GRP stress differs from that obtained following heat shock. In the case of heat shock, not all HSP-thermotolerance-inducing heat shocks induce a drug resistance (unpubl. data). What has been recognized by others (Osborne and MacKillop 1987) and is readily visible (by fluorescence microscopy) is the dramatic decline in adriamycin concentration in moderately to severely heated cells, indicating that resistance results largely from a reduction in the cellular drug concentration. While the multidrug resistant gene has been shown to have heat shock elements (Chin et al. 1990), its expression after heating would not seem to be sufficiently rapid to account for the immediate drug resistance obtained. Other factors such as cell cycle perturbations may contribute to the effect. This phenomenon indicates that major cellular/membrane transport functions can be altered by heat shock and such alterations are of significance in hyperthermia/cancer biology.

In the case of GRP-related stresses, the relationship between drug resistance and GRP induction was shown to be independent of drug retention by the cell and was not directly related to cell cycle redistribution (at least initially), indicating that another protective mechanism was at work (Colofiore et al. 1982; Shen et al. 1989). Since adriamycin is considered to have several possible modes of action in the cell, we analyzed a variety of anti-cancer agents of differing and more specific mechanisms of action. This study revealed that GRP stresses induced substantial resistance to the topoisomerase II targeted drug, etoposide (Hughes et al. 1989). Topoisomerase II is an enzyme that covalently binds to DNA, reversibly cutting it and allowing strand passage (Ross 1985). Etoposide inhibits the rejoining of DNA strands, leading to irreversible DNA scission. Protection of DNA against etoposide in GRP-stressed cells was verified by alkaline elution studies which demonstrated a reduction in DNA strand breakage during stress (Shen et al. 1989). Cell fractionation studies of GRP-stressed cells indicated that a rapid and substantial loss of topoisomerase II occurred from nuclei following the application of GRP stress. Furthermore, topoisomerase II appeared to be specifically sensitive to GRP stress, in that it was the predominant nuclear protein affected: other proteins in the polyacrylamide gel remained unaltered (with the possible exception that some GRP inducers lead to a depletion of HSP110). These observations were specific for GRP inducing stresses: nonlethal heat shocks (6–12 h at 41 °C or 5 min at 45 °C plus 6 h at 37 °C) which induced HSPs and thermotolerance increased cellular topoisomerase II levels (unpubl. data), indicating that the two categories of stress exhibit opposing influences on this proliferation-associated enzyme. In addition, virtually identical conclusions concerning the effect of GRP-inducing stress on topoisomerase II levels were obtained through analysis of the temperature-sensitive mutant K12 Chinese hamster cell line which induces GRP at the nonpermissive temperature (Shen et al. 1989).

The corresponding induction of GRPs in K12 cells, that were selected for growth inhibition (Tenner et al. 1977) in conjunction with the above described nuclear changes observed during glucose starvation and other GRP stresses, suggests a connection between the induction of GRPs and the loss of topoisomerase II observed under stress conditions that represents one pathway leading to the movement of cells out of cycle into Go/Gl and perhaps also freezing them in cycle (S-phase). However, the cell cycle redistribution is not sufficiently rapid to explain the resistance to either strand breaks or cell survival. Curiously, glucose starvation has also been associated with growth inhibition and/or differentiation of at least two different cellular systems (Taub et al. 1989; Le Bivic et al. 1988). Indeed, transfer of either system to glucose-free media results in GRP induction, despite the replacement of glucose by galactose in one of these cases (J Black and J Subjeck unpubl.). It might be inferred that these cells correspondingly lose topoisomerase II, stop synthesizing DNA, and cease cycling as a result of transfer to glucose-free medium.

It is most interesting that two differentially regulated but homologous sets of stress genes exist. Different stress proteins appear to be differentially compartmentalized. The fact that stress proteins in the endoplasmic reticulum are not heat-inducible in higher organisms while apparently similar stress proteins in the cytoplasm are not inducible by glucose starvation, anoxia etc. suggests a differential compartmental sensitivity to different categories of stress. Moreover, as discussed above, different sensitivities of different cellular compartments are also reflected at the level of proliferation-associated parameters in the nucleus. While glucose-regulated stresses send cells into a nonproliferative state, release from these conditions results in reinitiation of proliferation. In some instances the latter has been associated with the induction of heat shock proteins, e.g., anoxia and reoxygenation, glucose starvation, and addition of glucose (Subjeck and Shyy 1986; Whelen and Hightower 1985). Perhaps the later induction of HSPs may reflect the movement of an enriched Gl population of cells into S-phase following the release from GRP stress (Wu and Morimoto 1985). In any case, these and related changes and responses may be expected to be important factors in determining drug and heat resistance, proliferative status etc., with implications in many areas of cell physiology and tumor biology.

References

Arrigo AP, Suhan JP, Welch WJ (1988) Dynamic changes in the structure and intracellular locale of the mammalian low-molecular-weight heat shock protein. Mol Cell Biol 8:5059–5071

Black AR, Subjeck JR (1986) Correlation in the recovery of normal protein synthesis and the development of thermotolerance. J Cell Biol 103:188a

Black AR, Subjeck JR (1989) Involvement of rRNA, synthesis in enhanced survival and recovery of protein synthesis seen in thermotolerance. J Cell Physiol 138:439–449

Bole DG, Hendershot LM, Kearney JF (1986) Posttranslational association of immunoglobulin heavy chains in nonsecreting and secreting hybridomas. J Cell Biol 102:1558–1566

Born R, Eichholtz-Wirth H (1981) Effect of different physiological conditions on the action of adriamycin on Chinese hamster cells in vitro. Br J Cancer 44:241–246

Brugge JS, Erikson E, Erikson RL (1981) The specific interaction of the Rous sarcoma virus transforming protein, pp60^{src}, with two cellular proteins. Cell 25:363–372

Burdon RH (1986) Heat shock and the heat shock proteins. Biochem J 240:313–324

Caizergues-Ferrer M, Bouche G, Amalric F (1980) Effects of heat shock on RNA polymerase activities in Chinese hamster ovary cells. Biochem Biophys Res Commun 97:538–545

Chappel TG, Welch WJ, Schlossman DM, Polter KB, Schlesinger MJ, Rothman JE (1986) Uncoating ATPase is a member of the 70 kDa family of stress proteins. Cell 45:3–13

Chétien P, Landry J (1988) Enhanced constitutive expression of the 27-kDa heat shock proteins in heat-resistant variants from Chinese hamster cells. J Cell Physiol 137:157–166

Chin KV, Tanaka S, Darlington G, Pastan I, Gottesman M (1990) Heat shock and arsenite increase expression of the multidrug resistance gene in human renal carcinoma cells. J Biol Chem 265: 221–226

Chirico WJ, Waters MC, Blobel G (1988), 70 k heat shock related proteins stimulate protein translocation into microsomes. Nature (Lond) 332:805–810

Colofiore JR, Ara G, Berry D, Belli JA (1982) Enhanced survival of adriamycin-treated Chinese hamster cells by 2-deoxy-D-glucose and 2, 4-dinitrophenol. Cancer Res 42:3934–3940

Deshaies RJ, Koch BD, Werner-Washburne M, Craig EA, Schekman R (1988) A sub-family of stress proteins facilitates translocation of secretory and mitochondrial precursor polypeptides. Nature (Lond) 332:800–805

Duncan RF, Hershey JWB (1989) Protein synthesis and protein phosphorylation during heat stress, recovery, and adaptation. J Cell Biol 109:1467–1481

Easton DP, Rutledge PS, Spotila JR (1987) Heat shock protein induction and induced thermal tolerance are independent in adult salamanders. J Exp Zool 241:263–267

Fuhr JE (1974) Effect of hyperthermia on protein biosynthesis in L5178Y murine leukemic lymphoblasts. J Cell Physiol 84:365–372

Gerner EW, Schneider MJ (1975) Induced thermal resistance in HeLa cells. Nature (Lond) 256:500–502

Goldstein ES, Penman S (1973) Regulation of protein synthesis in mammalian cells. V. Further studies on the effect of actinomycin D on translation control in HeLa cells. J Mol Biol 80:243–254

Hahn GM, Strande DP (1976) Cytotoxic effects of hyperthermia and Adriamycin on Chinese hamster cells. J Natl Cancer Inst 57:1063–1067

Henle KJ, Leeper DB (1976) Interaction of hyperthermia and radiation of CHO cells recovery kinetics. Radiat Res 66:505–518

Henle KJ, Leeper DB (1979) Effects of hyperthermia (45°) on macromolecular synthesis in Chinese hamster ovary cells. Cancer Res 39:2665–2574

Henle KJ, Leeper DB (1982) Modification of the heat response and thermotolerance by cycloheximide, hydroxyurea, and lucanthone in CHO cells. Radiat Res 90:339–347

Hinds PW, Finlay CA, Frey AB, Levine AJ (1987) Immunological evidence for the association of p53 with a heat shock protein, hsc70, in p53-plus-ras-transformed cell lines. Mol Cell Biol 7:2863–2869

Hughes, CS, Shen JW, Subjeck JR (1989) Resistance to etoposide induced by three glucose regulated proteins in Chinese hamster ovary cells. Cancer Res 49:4452–4454

Hunt C, Morimoto RI (1985) Conserved features of eukaryotic hsp70 genes revealed by comparison with the nucleotide sequence of human hsp70. Proc Natl Acad Sci USA 82:6455–6459

Joab I, Radayi C, Renoir M, Buchou T, Catulli MG (1984) Common non-hormone binding component in non-transformed chick oviduct receptors for four steroid hormones. Nature (Lond) 308:850–853

Johnston RN, Kucey BL (1988) Competitive inhibition of HSP70 gene expression causes thermosensitivity. Science 242:1551–1554

Laszlo A (1989) The relationship of heat-shock proteins, thermotolerance, and protein synthesis. Exp Cell Res 178:401–414

Laszlo A, Li GC (1985) Heat resistant variants of Chinese hamster fibroblasts altered in expression of heat shock protein. Proc Natl Acad Sci USA 82:8029–8033

Le Bivic A, Hirn M, Reggio H (1988) HT-29 cells are an in vitro model for the generation of cell polarity in epithelia during embryonic differentiation. Proc Natl Acad Sci USA 85:136–140

Lee YJ, Dewey WC, Li GC (1987) Protection of Chinese hamster ovary cells from heat killing by treatment with cycloheximide or puromycin; involvement of hsp's? Radiat Res 111:237–253

Lewis MJ, Pelham HRB (1985) Involvement of ATP in the nuclear and nucleolar functions of the 70-kD heat shock protein. EMBO J 4:3137–3143

Li GC (1983) Induction of thermotolerance and enhanced heat shock protein synthesis in Chinese hamster fibroblasts by sodium arsenite and ethanol. J Cell Physiol 122:91–97

Li GC (1985) Elevated levels of 70,000 dalton heat shock protein in transiently thermotolerant Chinese hamster fibroblasts and their stable heat resistant variants. Int J Radiol Oncol Biol Phys 11:165–177

Li GC, Hahn GM (1978) Ethanol induced tolerance to heat and to adriamycin. Nature (Lond) 274:699–701

Li GC, Laszlo A (1985) Amino acid analogs, while inducing heat shock proteins, sensitize CHO cells to thermal damage. J Cell Physiol 122:91–97

Lindquist S, Craig EA (1988) The heat shock proteins. Annu Rev Genet 22:63–77

Mazzarella RA, Green M (1987) ERp99, an abundant conserved glycoprotein of the endoplasmic reticulum, is homologous to the 90 kDa heat shock protein (hsp90) and the 94 kDa glucose-regulated protein (grp94). J Biol Chem 262:8875–8883

Missen LA, Welch WJ (1988) Characterization of the thermotolerant cell. I Effect on protein synthesis activity and regulation of heat shock protein 70 expression. J Cell Biol 106:1105–1116

Munro S, Pelham H (1986) An hsp70-like protein in the ER: identity with the 78 kd glucose-regulated protein and immunoglobulin heavy chain binding protein. Cell 46:291–300

Oppermann H, Levinson W, Bishop HM (1981) A cellular protein that associates with the transforming protein of Rous sarcoma virus is also a heat shock protein. Proc Natl Acad Sci USA 78:1067–1071

Osborne EJ, MacKillop WJ (1987) The effect of exposure to elevated temperatures on the membrane permeability to adriamycin in Chinese hamster ovary cells in vitro. Cancer Lett 37:213–224

Pelham HRB (1984) Hsp70 accelerates the recovery of nucleolar morphology after heat shock. EMBO J 3:3095–3100

Perry RP, Kelly DE (1970) Inhibition of RNA synthesis by actinomycin D: characteristic dose-response of different RNA species. J Cell Physiol 76:127–140

Petersen NS, Mitchel HK (1981) Recovery of protein synthesis after heat shock: Prior heat treatment affects the ability of cells to translate mRNA. Proc Natl Acad Sci USA 78:1708–1711

Pouyssegur J, Shiu RPC, Pastan I (1977) Induction of two transformation-sensitive membrane polypeptides in normal fibroblasts by a block in glycoprotein synthesis of glucose deprivation. Cell 11:941–947

Riabowol KT, Mizzen LA, Welch WJ (1988) Heat shock is lethal to fibroblasts microinjected with antibodies against hsp70. Science 242:433–436

Rose DW, Wettenhall REH, Kudlicki W, Kramer G, Hardesty B (1987) The 90 kilodalton peptide of the heme-regulated eIf2α kinase has sequence similarity with the 90 kilodalton heat shock protein. Biochemistry 26:6583–6587

Rose DW, Welch WJ, Kramer G, Hardesty B (1989a) Possible involvement of the 90-kDa heat shock protein in the regulation of protein synthesis. J Biol Chem 264:6239–6244

Rose MD, Misra LM, Vogel JP (1989b) KAR2, a karyogamy gene, is the yeast homolog of the mammalian BiP/GRP78 gene. Cell 57:1211–1221

Ross W (1985) DNA topoisomerases as targets for cancer therapy. Biochem Pharmacol 34:4191–4195

Roti-Roti JL, Laszlo A (1988) The effects of hyperthermia in cellular macromolecules. Hyperthermia Oncol 1:13–56

Sadis S, Hickey E, Weber LA (1988) Effects of heat shock on RNA metabolism in HeLa Cells. J Cell Physiol 135:377–386

Sanchez ER, Meshinchi s, Tienrungroj W, Schlesinger MJ, Toff DO, Prott WB (1987) Relationship of the 90 kDa murine heat shock protein to the untransformed and transformed states of the L cell glucocorticoid receptor. J Biol Chem 262:6986–6991

Sciandra JJ, Subjeck JR (1983) The effects of glucose on protein synthesis and thermosensitivity in Chinese hamster ovary cells. J Biol Chem 258:12091–12093

Sciandra JJ, Subjeck JR (1984) Heat shock proteins and protection of proliferation and translation in mammalian cells. Cancer Res 44:5288–5294

Shen J, Hughes C, Chao C, Cai J, Bartels C, Gessner T, Subjeck J (1987) Coinduction of glucose-regulated proteins and doxorubicin resistance in Chinese hamster cells. Proc Natl Acad Sci USA 84:3278–3282

Shen JW, Subjeck JR, Lock RB, Ross W (1989) Depletion of topoisomerase II from isolated nuclei during a glucose regulated stress response. Mol Cell Biol 9:3284–3291

Sherwood SW, Daggnet AS, Shimke RT (1987) Interaction of hyperthermia and metabolic inhibitors on the induction of chromosome damage in Chinese hamster ovary cells. Cancer Res 47:3584–3588

Sorger PK, Pelham HRB (1987) The glucose regulated protein grp94 is related to the heat shock protein hsp90. J Mol Biol 194:341–344

Stevenson MA, Calderwood SK, Hahn GM (1987) Effect of hyperthermia (45 °C) on calcium flux in Chinese hamster ovary HA-1 fibroblasts and its potential role in cytotoxicity and heat resistance. Cancer Res 47:3712–3717

Subjeck JR, Shyy T-T (1986) Stress protein systems of mammalian cells. Am J Physiol 250 (Cell Physiol 19):C1–C17

Subjeck JR, Sciandra JJ, Johnson JR (1982) Heat shock proteins and thermotolerance: A comparison of induction kinetics. Br J Radiol 55:127–131

Subjeck JR, Shyy T-T, Shen JW, Johnson RJ (1983) Association between the mammalian 110,000 dalton heat shock protein and nucleoli. J Cell Biol 97:1389–1295

Taub ML, Syracuse JA, Cai JW, Fiorella P, Subjeck JR (1989) Glucose deprivation results in the induction of glucose-regulated proteins and domes in MDCK monolayers in hormonally defined serum free medium. Exp Cell Res 182:105–113

Tenner A, Zeig J, Scheffler IE (1977) Glycoprotein synthesis in a temperature-sensitive Chinese hamster cell cycle mutant. J Cell Physiol 90:145–160

Tomosovic (1989) Functional aspects of the mammalian heat-stress protein response. Life Chem Rep 7:33–63

Ungewickel E (1985) The 70-kd mammalian heat shock proteins are structurally and functionally related to the uncoating protein that releases clathrin triskelia from coated vesicles. EMBO J 4:3385–3391

Watowich S, Morimoto RI (1988) Complex regulation of heat shock- and glucose-responsive genes in human cells. Mol Cell Biol 8:393–405

Welch WJ, Feramisco JR (1985) Rapid purification of mammalian 70,000-dalton stress proteins: affinity of the proteins for nucleotides. Mol Cell Biol 5:1229–1237

Welch WJ, Suhan JP (1985) Morphological studies of the mammalian stress response: characterization of changes in cytoplasmic organelles, cytoskeleton and nucleoli, and appearance of intranuclear actin filaments in fat fibroblasts after heat shock. J Cell Biol 101:1198–1211

Whelen SA, Hightower LE (1985) Differential induction of glucose-regulated and heat shock proteins: effects of pH and sulfhydryl-reducing agents on chicken embryo cells. J Cell Physiol 125:251–258

Wu BJ, Morimoto RI (1985) Transcription of the human hsp70 gene is induced by serum stimulation. Proc Natl Acad Sci USA 82:6070–6074

Yaraha I, Iida H, Koyasu S (1986) A heat shock-resistant variant of Chinese hamster cell line constitutively expresses heat shock protein of M 90,000 at high level. Cell Struct Funct 11:65–73

Zhang JR, Shen JW, Subjeck JR (1990) Chronic anoxia leads to a complex of glucose regulated proteins. Proc 38th Ann Meet Radiat Res Soc, New Orleans, Louisiana, April 7, 1990 (Abstr)

Ziemiecki A, Catelli M-G, Joab I, Litwack G (1986) Association of the heat shock protein hsp90 with steroid hormone receptors and tyrosine kinase oncogene products. Biochem Biophys Res Commun 138:1298–1307

Subject Index